My revision notes

OCR GCSE (9–1)

GEOGRAPHY A

Rebecca Blackshaw
Jo Payne
Simon Ross

HODDER
EDUCATION
AN HACHETTE UK COMPANY

The Publishers would like to thank the following for permission to reproduce copyright material.

Photo credits

p.21 © Graham M. Lawrence / Alamy Stock Photo; **p.23** © Ian Ward (friend of author Alan Parkinon); **p.24** © Robert Estall photo agency / Alamy Stock Photo; **p.36** © Commission Air / Alamy Stock Photo; **p.53** © SWNS / Alamy Stock Photo; p.84 © World Mapper Ltd; **p.101** © Joerg Boethling / Alamy Stock Photo; **p.106** © Stefan Christmann/Corbis Documentary/Getty Images; **p.117** ©Rex Wholster - stock. adobe.com; **p.118** © National Geographic Creative / Alamy Stock Photo.

Acknowledgements

Every effort has been made to trace all copyright holders, but if any have been inadvertently overlooked, the Publishers will be pleased to make the necessary arrangements at the first opportunity.

Although every effort has been made to ensure that website addresses are correct at time of going to press, Hodder Education cannot be held responsible for the content of any website mentioned in this book. It is sometimes possible to find a relocated web page by typing in the address of the home page for a website in the URL window of your browser.

Hachette UK's policy is to use papers that are natural, renewable and recyclable products and made from wood grown in sustainable forests. The logging and manufacturing processes are expected to conform to the environmental regulations of the country of origin.

Orders: please contact Bookpoint Ltd, 130 Park Drive, Milton Park, Abingdon, Oxon OX14 4SE. Telephone: (44) 01235 827720. Fax: (44) 01235 400401. Email education@bookpoint.co.uk Lines are open from 9 a.m. to 5 p.m., Monday to Saturday, with a 24-hour message answering service. You can also order through our website: www.hoddereducation.com

ISBN: 978 1 5104 1893 6

© Rebecca Blackshaw, Jo Payne and Simon Ross 2018

First published in 2018 by
Hodder Education,
An Hachette UK Company
Carmelite House
50 Victoria Embankment
London EC4Y 0DZ
www.hoddereducation.co.uk

Impression number   10 9 8 7 6 5 4 3 2 1

Year   2022 2021 2020 2019 2018

Cover photo © iStock/Thinkstock/Getty Images

Illustrations by Integra and Barking Dog Art

Typeset by Integra Software Services Pvt. Ltd., Pondicherry, India

Printed in Spain

A catalogue record for this title is available from the British Library.

Get the most from this book

Everyone has to decide his or her own revision strategy, but it is essential to review your work, learn it and test your understanding. These Revision Notes will help you to do that in a planned way, topic by topic. Use this book as the cornerstone of your revision and don't hesitate to write in it – personalise your notes and check your progress by ticking off each section as you revise.

Tick to track your progress

Use the revision planner on pages 4 and 5 to plan your revision, topic by topic. Tick each box when you have:

● revised and understood a topic
● tested yourself
● practised the exam questions and gone online to check your answers.

You can also keep track of your revision by ticking off each topic heading in the book. You may find it helpful to add your own notes as you work through each topic.

Features to help you succeed

Exam tips

Expert tips are given throughout the book to help you polish your exam technique in order to maximise your chances of doing well in the exam.

Now test yourself

These short, knowledge-based questions provide the first step in testing your learning. Answers can be found online at **www.hoddereducation.co.uk/myrevisionnotes**

Definitions and key words

Clear, concise definitions of essential key terms are provided where they first appear in the main text.

Revision activities

These activities will help you to understand each topic in an interactive way.

Exam practice

Practice exam questions are provided for each topic. Use them to consolidate your revision and practise your exam skills.

Online

Go online to check your answers to the exam questions at **www.hoddereducation.co.uk/myrevisionnotes**

My revision planner

REVISED   TESTED   EXAM READY

Now test yourself and exam practice answers at **www.hoddereducation.co.uk/myrevisionnotes**

Answers to 'Now test yourself' and exam practice questions at www.hoddereducation.co.uk/myrevisionnotes

Countdown to my exams

6–8 weeks to go

- Start by looking at the specification — make sure you know exactly what material you need to revise and the style of the examination. Use the revision planner on pages 4 and 5 to familiarise yourself with the topics.
- Organise your notes, making sure you have covered everything on the specification. The revision planner will help you to group your notes into topics.
- Work out a realistic revision plan that will allow you time for relaxation. Set aside days and times for all the subjects that you need to study, and stick to your timetable.
- Set yourself sensible targets. Break your revision down into focused sessions of around 40 minutes, divided by breaks. These Revision Notes organise the basic facts into short, memorable sections to make revising easier.

REVISED ☐

2–6 weeks to go

- Read through the relevant sections of this book and refer to the exam tips and key terms. Tick off the topics as you feel confident about them. Highlight those topics you find difficult and look at them again in detail.
- Test your understanding of each topic by working through the 'Now test yourself' questions in the book. Look up the answers online at **www.therevisionbutton.co.uk/myrevisionnotes**.
- Make a note of any problem areas as you revise, and ask your teacher to go over these in class.
- Look at sample and/or past papers. They are one of the best ways to revise and practise your exam skills.
- Write or prepare planned answers to the exam practice questions provided in this book. Check your answers online at **www.therevisionbutton.co.uk/myrevisionnotes**.
- Use the revision activities to try out different revision methods. For example, you can make notes using mind maps, spider diagrams or flash cards.
- Track your progress using the revision planner and give yourself a reward when you have achieved your target.

REVISED ☐

One week to go

- Try to fit in at least one more timed practice of an entire past paper and seek feedback from your teacher, comparing your work closely with the mark scheme.
- Check the revision planner to make sure you haven't missed out any topics. Brush up on any areas of difficulty by talking them over with a friend or getting help from your teacher.
- Attend any revision classes put on by your teacher. Remember, he or she is an expert at preparing people for examinations.

REVISED ☐

The day before the examination

- Flick through these Revision Notes for useful reminders, for example the exam tips and key terms.
- Check the time and place of your examination.
- Make sure you have everything you need — extra pens and pencils, tissues, a watch, bottled water, sweets.
- Allow some time to relax and have an early night to ensure you are fresh and alert for the examinations.

REVISED ☐

My exams

Paper 1 Living in the UK today
Date: ....

Time: ....

Location: ....

Paper 2 The world around us
Date: ....

Time: ....

Location: ....

Paper 3 Geographical skills
Date: ....

Time: ....

Location: ....

1 Landscapes of the UK

The physical landscapes of the UK

How are landscapes distributed in the UK?

REVISED

The UK enjoys a huge variety of distinctive physical landscapes, including spectacular mountains (such as the Lake District in northern England and the Cairngorms in Scotland), rolling hills and valleys (much of southern England) and flat wetlands (such as the Somerset Levels).

Figure 1 is an atlas map showing the distribution of uplands and lowlands in the UK. Notice that most of the mountainous uplands in northern England, Wales and Scotland were glaciated in the past when the climate was very much colder than it is today. Immense glaciers carved spectacular valleys and created the dramatic landscapes that we associate with these mountainous areas.

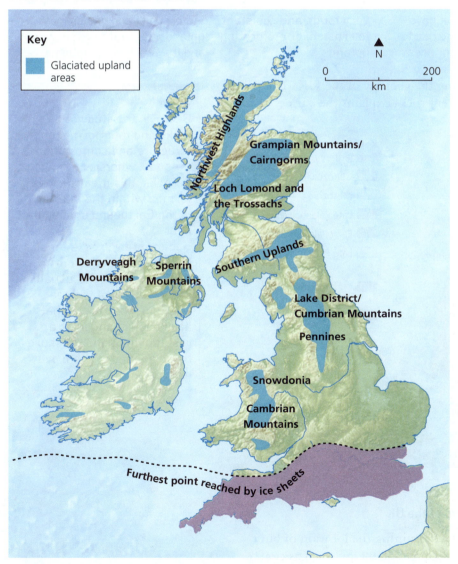

Figure 1 The distribution of upland, lowland and glaciated areas in the UK

The characteristics of landscapes in the UK

Geology

Geology is all about the rocks beneath our feet. It is one of the most important factors affecting the physical and human landscapes of the UK. It is possible to identify three types of rock: igneous, sedimentary and metamorphic.

Figure 2 Formation, characteristics and examples of the three rock types

Rock type	Formation	Characteristics	Examples
Igneous	Magma that has cooled either on the ground surface (extrusive) when a volcano erupts or below the ground (intrusive)	Tough and resistant to erosion; igneous rocks often form uplands	Dartmoor (granite), Northern Ireland (basalt), Cuillin Hills (gabbro)
Sedimentary	Rocks formed from the accumulation and compaction of sediment, usually in the ocean	Variable resistance to erosion; chalk and limestone are resistant and will form uplands, whereas weaker clays and sands form lowlands	Chalk ridges and escarpments (for example, Lincolnshire, the Chilterns and the South Downs), limestone (for example, the Pennines), and sands and gravels (lowlands in southern England)
Metamorphic	Existing rocks that have undergone change due to extreme heating or pressure	Very tough and resistant to erosion, often forming uplands	Slate, schist and gneiss form uplands (for example, Snowdonia and the Scottish mountains)

Figure 3 is a simplified map showing the geology of the UK. The great range of colours representing different rocks explains the variety of UK landscapes.

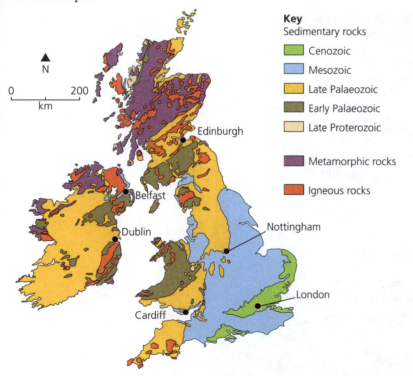

Figure 3 Simplified geological map of the UK

Key

Sedimentary rocks
- Cenozoic
- Mesozoic
- Late Palaeozoic
- Early Palaeozoic
- Late Proterozoic

- Metamorphic rocks
- Igneous rocks

Key terms

Geology: the study of rocks and their formation, structure and composition

Igneous: when referring to rocks, this means rocks formed within the interior of the Earth, and shaped by heat

Sedimentary: rocks that have been produced from layers of sediment, usually at the bottom of the sea

Metamorphic: rocks that have been changed as a result of heat and pressure being applied to them over long periods of time

---

Geology has also had an influence determining the location of built landscapes. Some rocks are valuable sources of energy (such as coal) or contain raw materials (such as metal ores found in limestone). Where these rocks occur at or close to the surface, they have encouraged

Now test yourself and exam practice answers at www.hoddereducation.co.uk/myrevisionnotes

industrial location and **urbanisation**. In the north-east of England, the development of Middlesbrough was based on nearby mineral resources, which supplied the local chemical industry.

Climate

Climate is the long-term average weather conditions, calculated over a period of 30 years. It differs from the weather, which is the day-to-day condition of the atmosphere (temperature, humidity, **wind** speed, and so on).

The climate of the UK has a role to play in creating the UK's distinctive landscapes, but it is also affected *by* the landscapes – it is a two-way relationship.

The UK enjoys a maritime climate, with the prevailing (most common) winds blowing across the Atlantic from the south-west. This accounts for the generally high rainfall and moderate temperatures throughout the year.

- **Rainfall**: the uplands – particularly in the west – receive a high proportion of the rainfall, with the lowlands in the south and east tending to be drier. This is because the moist air from the Atlantic is forced to rise and cool over the western uplands, forming rain-bearing clouds. This type of rainfall is called relief rainfall. The drier regions to the east can be described as being in the '**rain shadow**'.
- **Temperatures**: temperatures tend to be lower in the uplands than in the lowlands, with frost and snow being common hazards in the winter. This is because temperature falls on average by 0.6 °C per 100 metres of altitude.

Climate affects physical processes such as **weathering** and erosion.
- The process of **freeze–thaw** is very active in upland areas, resulting in jagged rock surfaces and accumulations of scree on mountainsides.
- Rivers are fast flowing (due to high rainfall) and very erosive in uplands, carving deep V-shaped valleys.
- In the past, extreme cold in the uplands led to the formation of ice caps and glaciers, which carved spectacular landscapes.

Human activity

Human activity has transformed the landscape of the UK from a largely forested landscape at the end of the last ice age (about 10,000 years ago) to the present day's agricultural and urbanised landscape.
- **Uplands**: these areas are sparsely populated due to the harsh climate and steep relief. Human activity is limited to extensive sheep rearing and forestry. Reservoirs have been created in some areas to supply water and to generate hydroelectric power. In recent years, wind farms have been constructed in some upland areas, exploiting the strong winds.
- **Lowlands**: these areas are more densely populated due to the moderate climate and gentler relief. Commercial farming dominates the countryside and much of this landscape is urbanised or criss-crossed by transport and service infrastructure.

Exam practice

1 Outline the main characteristics of the built environment. [2]
2 Describe how human activities have affected the landscape of the UK. [4]
3 Explain how geology and climate have affected the upland landscapes of the UK. [6]

ONLINE ☐

Key term

Urbanisation: the process of towns and cities developing and becoming bigger as their population increases

Now test yourself

Describe the formation and landscape characteristics of igneous, sedimentary and metamorphic rocks.

Answers online

TESTED ☐

Key terms

Wind: the movement of air on a large scale over the Earth's surface

Rain shadow: an area or region behind a hill that has little rainfall because it is sheltered from rain-bearing winds

Weathering: the breakdown of material *in situ* by physical, chemical and biological processes; if movement is involved, this becomes erosion

Freeze–thaw cycle: the daily fluctuations of temperature either side of freezing point; when repeated they contribute to physical weathering

Revision activity

Construct a spider diagram to show how geology, climate and human activity affect UK landscapes (uplands and lowlands).

Geomorphic processes

What are the main geomorphic processes?

Geomorphic processes are responsible for shaping landscapes. They include weathering, mass movement, erosion, **transportation** and **deposition**.

Weathering

Weathering involves the decomposition or disintegration of rock in its original place at or close to the ground surface. There are two main types of weathering: **chemical weathering** and **mechanical (physical) weathering**.

Figure 1 Processes of chemical and mechanical weathering

Chemical weathering	Mechanical weathering
Carbonation: carbon dioxide dissolved in rainwater forms a weak carbonic acid; this reacts with calcium carbonate (limestone and chalk) to form calcium bicarbonate, which is soluble and can be carried away by water	**Freeze–thaw**: repeated cycles of freezing and thawing causes water trapped in rocks to expand and contract, eventually causing rock fragments to break away (Figure 2)
Hydrolysis: acidic rainwater reacts with feldspar in granite, turning it into clay and causing granite to crumble	**Salt weathering**: crystals of salt, often evaporated from seawater, grow in cracks and holes, expanding to cause rock fragments to flake away
Oxidation: oxygen dissolved in water reacts with iron-rich minerals, causing rocks to crumble	

A third type, **biological weathering**, involves living organisms such as nesting birds, burrowing animals and plant roots. Plants roots may expand in cracks, slowly prising rocks apart. Acids that promote chemical weathering may be active beneath **soils** and rotting vegetation.

Key terms

Biological weathering: weathering that results from the action of living organisms, such as plants or animals

Soil: the top layer of the earth in which plants grow; it contains organic and inorganic material

Now test yourself

1 What is the difference between mechanical and chemical weathering?
2 Outline the process of freeze–thaw weathering.
3 How does the action of plant roots cause weathering to rocks?

Answers online

Key terms

Geomorphic processes: processes that result in a change in the nature of landscapes

Transportation: the movement of eroded material

Deposition: the laying down of materials that have been transported and can create new landforms such as beaches

Chemical weathering: the decomposition of rocks involving a chemical change, usually resulting from acidic water

Mechanical (physical) weathering: the disintegration or break-up of rocks without any chemical change

Hydrolysis: chemical breakdown of a material due to interaction with water

Oxidation: a chemical reaction between a substance and the air; it can change its appearance or weaken it

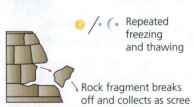

Figure 2 Freeze–thaw weathering

Mass movement

Mass movement is active at the coast, particularly where cliffs are undercut by the sea, making them unstable. It includes sliding and slumping, as well as falls (rockfalls) and flows (mudflows).

- **Sliding**: this involves rock or loose material sliding downhill along a slip plane, such as a **bedding plane**. Slides are often triggered by ground shaking (for example, an earthquake) or heavy rain.
- **Slumping**: this commonly involves the collapse of weak rock, such as sands and clays, often found at the coast. Slumping often results from heavy rainfall when the **sediments** become saturated and heavy.

Common forms of mass movement at the coast include:

- Rockfall: individual fragments or chunks of rock falling off a cliff face, often resulting from freeze–thaw weathering.
- Landslide: blocks of rock sliding rapidly downslope along a linear shear-plane, usually lubricated by water (Figure 3).
- Mudflow: saturated material (usually clay) flowing downhill, which may involve elements of sliding or slumping as well as flow.
- Rotational slip/slump: slumping of loose material often along a curved shear-plane, lubricated by water (Figure 4).

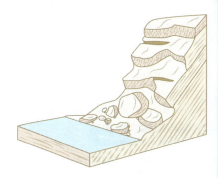

Figure 3 Landslide

Revision activity

Make a copy of Figure 3 and add labels to describe the causes and characteristics of sliding.

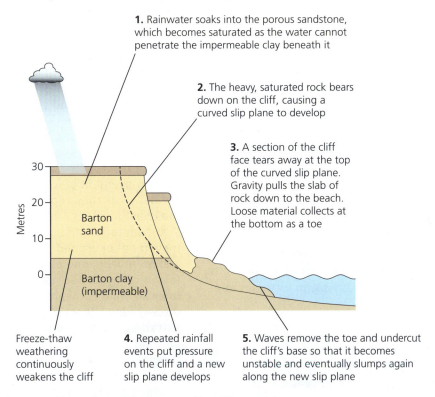

1. Rainwater soaks into the porous sandstone, which becomes saturated as the water cannot penetrate the impermeable clay beneath it

2. The heavy, saturated rock bears down on the cliff, causing a curved slip plane to develop

3. A section of the cliff face tears away at the top of the curved slip plane. Gravity pulls the slab of rock down to the beach. Loose material collects at the bottom as a toe

Freeze-thaw weathering continuously weakens the cliff

4. Repeated rainfall events put pressure on the cliff and a new slip plane develops

5. Waves remove the toe and undercut the cliff's base so that it becomes unstable and eventually slumps again along the new slip plane

Figure 4 Slumping at Barton on Sea, Hampshire

Now test yourself

TESTED ☐

Use Figure 4 to describe the causes and characteristics of slumping.

Answers online

Erosion

Erosion involves the wearing away and removal of material by a moving force, such as a breaking wave. The main processes of erosion are **abrasion**, **hydraulic action**, **attrition** and **solution**.

- **Abrasion**: this is the 'sandpapering' effect as loose rock particles carried by the water scrape against solid bedrock. It can also involve loose particles being flung against a sea cliff or river bank by the water – a process sometimes referred to as corrasion.
- **Hydraulic action**: this involves the sheer power of the water, often compressing air into cracks in sea cliffs or river banks, causing rocks to break away.
- **Attrition**: erosion caused when rocks and boulders transported by **waves** bump into each other and break up into smaller pieces. Over time the rocks become smaller and more rounded.
- **Solution**: the dissolving of soluble rocks, such as chalk and limestone.

Transportation

Transportation involves the movement of eroded sediment from one place to another. It commonly involves the following processes:

- **Traction**: large particles rolling along the seabed.
- **Saltation**: a bouncing or hopping motion by pebbles too heavy to be suspended.
- **Suspension**: particles suspended within the water.
- **Solution**: chemicals dissolved in the water.

> **Revision activity**
>
> Draw a simple diagram to show the four processes of transportation. Set your diagram in a river or at the coast.

Deposition

Deposition occurs when material being transported is dropped due to a reduction in energy. This typically occurs in areas of low energy, where velocity is reduced and sediment can no longer be transported. At the coast, deposition is common in bays or in areas sheltered by bars and spits. In rivers, deposition is common close to the river banks, in estuaries and at the inside bend of meanders.

Exam practice

1 Describe the process of mechanical weathering. [2]
2 Explain the conditions under which hydraulic action will be an important process of coastal erosion. [4]
3 Explain where and why sediment is deposited in rivers and at the coast. [6]

ONLINE

Key terms

Erosion: the wearing away and removal of material by a moving force

Abrasion (or corrasion): the scraping, scouring or rubbing action of materials being carried by moving features such as rivers, glaciers or waves, which erode rocks

Hydraulic action: an erosive process that involves the pressure of water hitting a surface, compressing air in any cavities that exist, and resulting in the removal of rock fragments over time

Attrition: a reduction in the size of material

Solution (or corrosion): a type of erosion that involves rock being chemically changed such that it is taken into solution and removed, e.g. the action of acidic water on limestone

Waves: elliptical or circular movement of the sea surface that is translated into a movement of water up the beach as they approach the coastline

Transportation: the movement of eroded sediment from one place to another, you also need to know what is meant by the key terms **traction**, **saltation**, **suspension** and **solution**

Deposition: when material being transported is dropped due to a reduction in energy

River landforms
The formation of river landforms

V-shaped valleys

A **V-shaped valley**, as the term implies, is a steep-sided, narrow river valley that takes the form of a V-shape in its cross profile. It is formed by river erosion as the river cuts down vertically into the mountainous landscape. Weathering and mass movement on the valley sides are responsible for broadening the top of the valley profile. Interlocking spurs – 'fingers' of land jutting into the valley – are common landforms in the upper course of a river (Figure 1).

> **Key term**
>
> **V-shaped valley**: a steep-sided, narrow river valley that takes the form of a V-shape in its cross profile

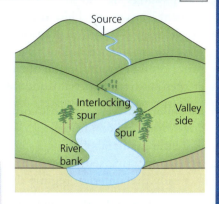

Figure 1 Interlocking spurs

Waterfalls

A **waterfall** is typically formed where fast-flowing water plummets over a vertical cliff – often a considerable drop – into a deep plunge pool below (Figure 2). Hydraulic action and abrasion are mainly responsible for eroding the plunge pool, which can be several metres deep in the centre.

Waterfalls most commonly form when a river flows over a hard, resistant band of rock. Unable to erode the tougher rock, a 'step' is formed in the long profile of the river.

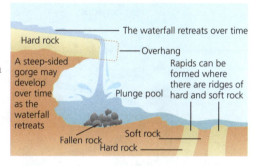

Figure 2 The formation of a waterfall

> **Key terms**
>
> **Waterfall**: steep fall of river water where its course crosses between different rock types, resulting in different rates of erosion
> **Gorge**: narrow valley between hills or mountains

Gorges

A **gorge** is commonly formed by the upstream retreat of a waterfall (Figure 3).

- Erosion (primarily hydraulic action and abrasion) undercuts the hard rock forming the waterfall to create an overhang.
- Eventually this overhang collapses into the plunge pool causing the waterfall to retreat upstream.
- Over many thousands of years, this process of undercutting and collapse continues and a gorge is formed immediately downstream.

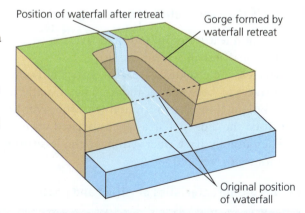

Figure 3 The formation of a gorge

> **Key term**
>
> **Gorge**: a narrow, steep sided valley, often formed as a waterfall retreats upstream

Meanders

Meanders are commonly found in a river's middle and lower course, where they can form extensive and elaborate bends. The fastest flowing water swings around the outside bend of a meander, eroding the banks to form a river cliff. Here the water is deep. On the inside bend, where the velocity is lower, deposition occurs, forming a slip-off slope. In this way, the meander develops an asymmetrical cross profile (Figure 4). Over time, lateral erosion on the outside bend widens the river valley and creates an extensive, flat **floodplain**.

Meandering rivers are most commonly associated with the following environmental conditions:

● gentle gradients
● relatively fine sediments
● steady precipitation regime throughout the year.

This explains why they are by far the most common river pattern in the UK and, indeed, throughout the world.

> **Key terms**
>
> **Meander:** a bend in a river that results from the flow of water along it
>
> **Floodplain:** the flat area of land either side of a river channel forming the valley floor, which may be flooded

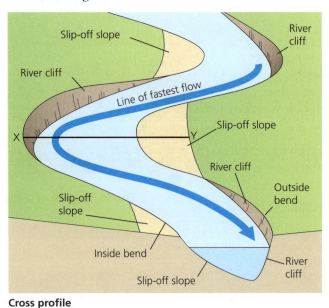
Cross profile

Figure 4 Characteristics of a river meander

A common feature of meandering rivers is an alternating pattern of shallows (riffles) and deeps (pools).

● **Riffles:** these shallow areas are associated with the straighter sections of rivers in-between meanders. They usually have rocky beds and turbulent flow due to friction with the river bed.
● **Pools:** these deeper areas are associated with the meander bends. They usually have finer sediment and less turbulent flow due to the smoother river bed.

> **Now test yourself**
>
> Draw a sketch cross profile from X to Y on Figure 4 and explain the formation of the features of erosion and deposition.
>
> **Answers online**
>
> TESTED ☐

Figure 5 Characteristics of riffles and pools

Rate of flow	Riffle	Pool
High flow (winter)	Greater friction in the shallower riffles results in slower, more turbulent flow	Water tends to flow faster through the deeper pools due to a reduction in friction with the bed and banks
Low flow (summer)	On entering a riffle, the reduction in channel size often results in slightly faster flow	The reduced volume of water tends to slow down on entering a deep pool, where the channel is larger

Ox-bow lakes

Ox-bow lakes are typically seen with rivers in their middle or lower course. They represent old meander bends that have been cut off by faster flowing water during times of flood (Figure 6). Gradually, without water flowing through them, they will become filled with silt and marshy vegetation will grow.

Stage 1

Narrow neck being eroded

- Neck of meander narrows due to lateral erosion on opposite sides of the meander bend.

Stage 2

New straight course

Redundant loop

- During high flow (flood) conditions, the meander neck is broken through.
- The river now adopts the shorter (steeper) route, bypassing the old meander.

Stage 3

River

Deposition helps infill and detach the loop

Oxbow lake

Marsh plants colonise, drying out area

- Deposition occurs at the edges of the new straight section, effectively cutting off the old meander.
- The old meander now forms an oxbow lake, separated from the main river.
- Gradually the oxbow lake silts-up to form marshland.

Figure 6 Formation of an ox-bow lake

> **Key term**
>
> **Ox-bow lake**: a horseshoe-shaped lake that forms when a meander is separated from the main river channel as a result of erosion

> **Now test yourself**
>
> Describe the formation of an ox-bow lake.
>
> **Answers online**
>
> TESTED ☐

Levees

Levees are raised river banks commonly found in the lower courses of rivers. They are formed during flood conditions when water flows over the river banks on to the surrounding floodplain (Figure 7) in a sequence of steps:

1 As water overtops the river banks, there is a sudden localised reduction in the velocity of the water, which had previously been flowing very fast along the river channel.

2 This causes sediment in suspension to be deposited at the river bank.

3 Coarse (heavier) sediment is deposited first and this traps the finer sediment.

4 With each successive flood, the deposited sediment raises the river banks by as much as a few metres.

Levees can be created artificially by people to contain water within a river channel to reduce the threat of flooding. In the USA, the term 'levee' is most commonly used for the artificial form.

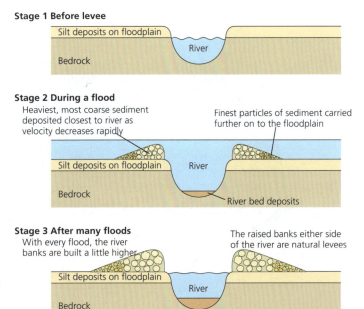

Stage 1 Before levee

Silt deposits on floodplain

River

Bedrock

Stage 2 During a flood

Heaviest, most coarse sediment deposited closest to river as velocity decreases rapidly

Finest particles of sediment carried further on to the floodplain

Silt deposits on floodplain

River

Bedrock

River bed deposits

Stage 3 After many floods

With every flood, the river banks are built a little higher

The raised banks either side of the river are natural levees

Silt deposits on floodplain

River

Bedrock

River bed builds up bed load deposits over time. This raises the level of the river so increases probability of flooding

Figure 7 Formation of a levee

> **Now test yourself**
>
> TESTED ☐
>
> Use a series of simple diagrams to explain the formation of a levee.
>
> **Answers online**

> **Key term**
>
> **Levee**: raised banks along a river

Floodplains

Floodplains are associated with rivers in their middle and lower course. They are extensive, flat areas of land mostly covered by grass. There may be some marshy areas close to the river, and also ox-bow lakes (Figure 8). As the name implies, they are formed during flood conditions and are periodically and quite naturally inundated by water.

1 During a flood, water containing large quantities of alluvium (river silt) pours out over the flat valley floor.
2 The water slowly soaks away, leaving the deposited sediment behind.
3 Over hundreds of years, repeated flooding forms a thick alluvial deposit that is fertile and often used for farming.

Floodplains become wider due to the lateral erosion of meanders.

1 When the outside bend of a meander meets the edge of the river valley, erosion will cut into it, thereby widening the valley at this point.
2 As meanders slowly migrate downstream, the entire length of the valley will eventually be widened.

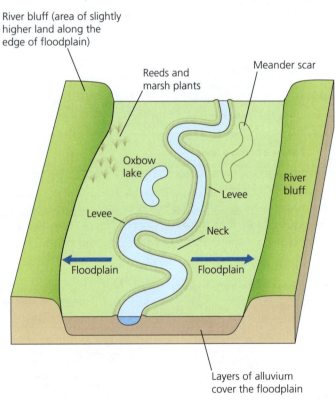

Figure 8 Characteristic features of a floodplain

Exam practice

1 Explain the formation of a V-shaped valley. [4]
2 Describe the characteristics of a waterfall. [4]
3 Explain how the processes of erosion and deposition are responsible for forming the characteristic features of a meander. [6]
4 Explain the formation of levees and floodplains. [6]

ONLINE

Exam tip

In question 2, focus on the *characteristics* only. You do not need to describe the *formation* of a waterfall and will get no marks for doing so. You could use a labelled diagram to support your answer.

Coastal landforms

The formation of coastal landforms

Headlands and bays

Headlands and **bays** are characteristic features of a discordant coastline where rocks of different hardness (resistance to erosion) are exposed at the coast. Headlands form resistant promontories jutting out into the sea. They are separated by bays of less-resistant rock where the land has been eroded back by the sea.

Look at Figure 1. Notice that coastal erosion processes erode away the weaker rocks more readily than the harder rocks to form a sequence of alternating headlands and bays.

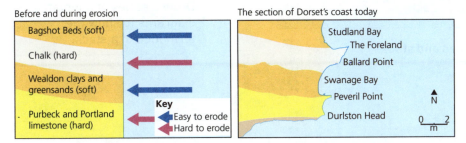

Figure 1 Formation of headlands and bays on the Dorset coast

Erosion at the headlands creates landforms such as cliffs and **wave-cut platforms**. Deposition occurs in the sheltered waters of the bay, forming a beach.

> **Revision activity**
>
> Create your own diagram to show the formation of headlands and bays. Indicate the concentration of erosion at the headland and deposition in the bay.

Caves, arches and stacks

Caves, **arches** and **stacks** are commonly formed at headlands where relatively tough rock juts out into the sea. They are formed in several steps:

1 Cracks in the cliff (**joints** or faults) are eroded and enlarged by hydraulic action to form a wave-cut notch at the base of the cliff. This is eroded further by hydraulic action and abrasion to form a cave.
2 The cave is eroded right through the headland to form an arch.
3 Over time, processes of erosion widen the arch and processes of weathering weaken its roof.
4 Eventually the roof collapses to form an isolated pillar of rock called a stack. Over time the stack will be eroded and will collapse to form a **stump**, which is only exposed at low tide.

> **Key terms**
>
> **Headland**: an area of land that extends out into the sea, usually higher than the surrounding land
>
> **Bay**: an area of the coast where the land curves inwards
>
> **Wave-cut platform/notch**: a flat area along the base of a cliff produced by the retreat of the cliff as a result of erosive processes
>
> **Cave**: a natural underground chamber or series of passages, especially with an opening to the surface; also refers to the extended cracks at the base of a cliff
>
> **Arch**: an arch-shaped structure formed as a result of natural processes within a rock feature such as a cliff
>
> **Stack**: a coastal feature that results from erosion; a section of headland that has become separated from the mainland and stands as a pillar of rock
>
> **Joints**: vertical cracks within a rock, such as limestone, which result from the natural shrinking of the rock over time as it was formed; these may form weaknesses allowing water to penetrate the rock
>
> **Stump**: a coastal feature that results from the collapse of a stack to form a protrusion of rock close to the sea surface

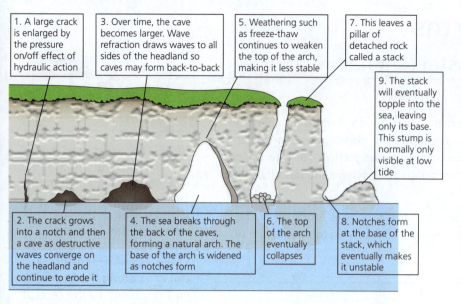

1. A large crack is enlarged by the pressure on/off effect of hydraulic action

2. The crack grows into a notch and then a cave as destructive waves converge on the headland and continue to erode it

3. Over time, the cave becomes larger. Wave refraction draws waves to all sides of the headland so caves may form back-to-back

4. The sea breaks through the back of the caves, forming a natural arch. The base of the arch is widened as notches form

5. Weathering such as freeze-thaw continues to weaken the top of the arch, making it less stable

6. The top of the arch eventually collapses

7. This leaves a pillar of detached rock called a stack

8. Notches form at the base of the stack, which eventually makes it unstable

9. The stack will eventually topple into the sea, leaving only its base. This stump is normally only visible at low tide

Figure 2 Formation of caves, arches and stacks

Now test yourself

TESTED ☐

Explain the formation of an arch.

Answers online

Beaches

A **beach** is a depositional landform made of sand or pebbles (shingle) extending from the low water line to the upper limit of storm waves. Beaches may exhibit a range of small landforms such as ridges called berms (formed by waves just above the high tide line, for example), ripples or shallow water-filled depressions called runnels. Figure 3 shows a typical beach profile.

- **Sandy beaches**: commonly formed in sheltered bays, associated with relatively low-energy constructive waves. Flat and extensive beaches are often backed by sand dunes.
- **Pebble beaches**: commonly associated with higher-energy coastlines where destructive waves remove finer sand, leaving behind coarser pebbles. Beaches tend to be steep and narrow with distinctive high tide berms.

Now test yourself

Suggest reasons why some stretches of coastline have sandy beaches.

Answers online

TESTED ☐

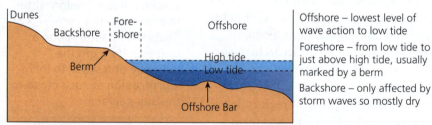

Dunes
Backshore
Fore-shore
Offshore
Berm
High tide
Low tide
Offshore Bar

Offshore – lowest level of wave action to low tide

Foreshore – from low tide to just above high tide, usually marked by a berm

Backshore – only affected by storm waves so mostly dry

Figure 3 Beach profile

Spits

A **spit** is a sand or shingle (pebble) ridge most commonly formed by **longshore drift** operating along a stretch of coastline (Figure 4). It is rather like a narrow beach extending out from the coast into the sea. At its end, where it is more exposed to variations in wind and waves, it tends to curve to form a hook or recurved tip. Some spits have sand dunes on them.

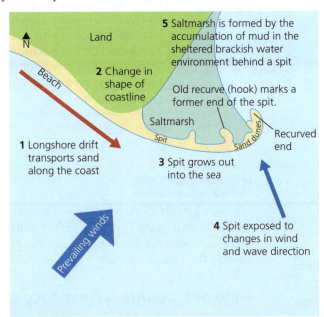

Figure 4 Formation of a spit

Key terms

Spit: a sand or shingle ridge most commonly formed by longshore drift operating along a stretch of coastline
Longshore drift: the movement of sediments along a stretch of coastline as a result of wave action
Drift: all sediment deposited by glaciers

Exam tip

Remember that a spit is land, so it lies above the high tide line. On an Ordinance Survey map, use the high tide line to trace the edges of a spit.

Revision activity

Make up your own version of Figure 4 with longshore drift operating from north to south along a stretch of coastline. Ensure that your diagram is fully labelled.

Exam tip

Make use of annotated diagrams when describing the characteristics or formation of coastal landforms.

Exam practice

1 Describe how rock type can lead to the formation of headlands and bays.   [4]
2 Explain the formation of caves, arches and stacks.   [6]
3 Explain how longshore drift leads to the formation of a spit.   [6]

ONLINE

Landscapes of the UK: case studies

Case study: A UK river basin

In this section you need to study **one** UK river basin. You should focus on the following aspects:

- geomorphic processes operating at different scales; how they are influenced by geology and climate
- landforms created by geomorphic processes
- the impact of human activity (including management) on geomorphic processes and the landscape.

Case study: River Wye

At over 210 kilometres in length, the River Wye is the fifth-longest river in the UK. From its source, high up in the Plynlimon Hills in central Wales, the River Wye flows roughly south-eastwards to join the River Severn at Chepstow. For much of its course the river flows through moorland and farmland. Considered to be a fairly natural river, there are few alterations resulting from human activity, for example, the construction of dams and reservoirs. However, water power was used in paper production in the past, which was heavily polluting.

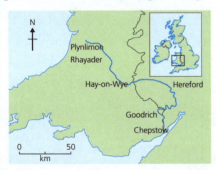

Figure 1 The course of the River Wye

Geomorphic processes

Geomorphic processes operate at different scales, both spatially (space) and temporally (time). The following list includes some of the geomorphic processes seen at the River Wye:

- Processes of river erosion, transportation and deposition are active along the course of the River Wye producing a range of distinctive landforms, such as a V-shaped valley, meanders, floodplains and levees.
- Weathering and mass movement are active, particularly in the river's upper course in the Plynlimon Hills. Compared with river processes, weathering and mass movement are more localised and take place at a smaller scale.

Revision activity

Draw a simple sketch to show the course of the River Wye (Figure 1). Use a series of detailed labels to identify some of the landforms formed along the river's course.

How are geomorphic processes affected by geology and climate?

Geology: For much of its upper course, the River Wye flows across **impermeable** shales and gritstones. This accounts for the large number of tributaries that join the river. High rates of flow occurring after rainfall events enable the river to carry out significant erosion, forming steep-sided, V-shaped valleys (the upper slopes are actively weathered and affected by mass movement processes), waterfalls and rapids.

Key term

Impermeable: a surface or substance that doesn't allow water to pass through it

Away from the Plynlimon Hills, geology continues to influence geomorphic processes and landform development:

- Near Rhayader, alternating bands of hard and soft rock result in a series of rapids, a very popular site for canoeists.
- To the south of Hereford, weak mudstones and sandstones have been easily eroded to form an extensive flat valley characterised by sweeping meanders.
- In the south, between Goodrich and Chepstow, the river cuts through tough carboniferous limestone, forming an impressive gorge known as the Wye Valley.

Climate

The average annual rainfall of the River Wye basin is 725 mm. In the Plynlimon Hills, the annual rainfall can exceed 2500 mm. Much of this rainfall occurs in the winter when there is little growing vegetation to absorb the surplus water. This leads to rapid river flows, high rates of erosion and the potential for flooding.

Winter temperatures can be low, particularly in the Plynlimon Hills. This results in active freeze–thaw weathering on the exposed river valley sides. Active weathering, together with the high rainfall, promotes the processes of mass movement, such as sliding and slumping.

River landforms

The River Wye exhibits landforms of both erosion and deposition.

- V-shaped valleys are found in the upper course of the River Wye and along its tributaries in the Plynlimon Hills.
- Waterfalls (for example, Cleddon Falls) are formed on some of the River Wye's tributaries; the stretch of river near Rhayader is known for its spectacular series of rapids.
- The Wye Valley, a steep-sided river gorge, extends between Goodrich and Chepstow.
- Sweeping meanders have formed on the flat lowland plains to the south of Hereford.
- Levees and floodplains formed by extensive alluvium deposition are found in the middle and lower courses of the river.

Impact of human activity and management

The Environment Agency describes the River Wye as being 'highly urbanised'. The river flows through several large settlements including Rhayader, Hay-on-Wye, Hereford and Chepstow. Over 200,000 people live in the Wye and Usk (its tributary) river valleys. Much of the valley is used for farming, particularly in the river's middle and lower courses. The Wye Valley is popular with tourists who enjoy the attractive landscape, wildlife and opportunities for adventure tourism, such as kayaking, canoeing and climbing.

Flooding is a serious issue, particularly in **urban** settlements close to the river. To reduce the flood risk in Hereford the following steps have been taken:

Key term
Urban (built): refers to areas that have been built by people; towns and cities

- Storage lakes (such as Letton Lake) have been constructed above the town to store surplus water.
- Parts of the floodplain above the town are deliberately allowed to flood, relieving the pressure downstream.
- In Hereford itself, flood walls have been constructed to protect some 200 properties in the Belmont area at a cost of over £5 million (Figure 2).

Human activity and river management have affected geomorphic processes in the following ways:

- Tree planting in the river's upper course helps to stabilise the slopes, reducing mass movement. This reduces the amount of sediment in the river, which can choke the channel and increase the risk of flooding. With less sediment in the river, there will be a reduction in deposition further downstream.
- Fewer flooding events mean that there is less sediment available to construct floodplains, and natural levees might not build up. Artificial levees will have to be constructed instead.
- River banks along the course of the river have been stabilised by planting vegetation, improving access for anglers and walkers. Planting trees can reduce the height of floods by twenty per cent as they increase the amount of water that can be stored. The management of river banks will affect rates of flow and river processes, both in towns and in the countryside.

Figure 2 Flood defences protect property in Hereford, December 2012

Case study: A UK coastal landscape

In this section you need to study **one** coastal landscape. You should focus on the following aspects:

● geomorphic processes operating at different scales; how they are influenced by geology and climate
● landforms created by geomorphic processes
● the impact of human activity (including management) on geomorphic processes and the landscape.

Case study: North Norfolk coast

The North Norfolk coast is located in East Anglia in the east of England (Figure 3). It is a low-lying stretch of coastline with a variety of habitats including broad sandy beaches, salt marshes, sand dunes and some stretches of cliffed coastline, particularly east of Cromer. The landscape is famous for its 'big skies', attracting tourists and artists.

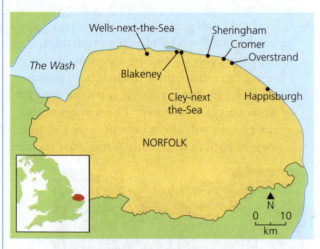

Figure 3 The North Norfolk coast

Geomorphic processes

Geomorphic processes operate at different scales, both spatially (space) and temporally (time). The following list includes a few of the geomorphic processes seen on the North Norfolk coast:

● The North Norfolk coast is exposed to powerful waves from the north-east, the direction of maximum fetch (over 4000 km). When the winds come from the north-east, waves are at their most powerful.
● Longshore drift predominantly operates from east to west along this stretch of coastline, although opposing currents transport sediment southwards along the east coast of England, around the Wash.
● Mechanical weathering processes, such as freeze–thaw and mass movement (particularly slumping), are very active on the cliffs in the east

of the region, particularly between Overstrand and Happisburgh. Unlike longshore drift, these processes are more localised and smaller scale.
● Coastal erosion is active in the east, with coastal deposition dominating in the west.

How are geomorphic processes affected by geology and climate?

Geology The entire region is underlain by the sedimentary rock chalk, which is exposed in places such as the base of the cliffs at Overstrand. Overlying the chalk is thick glacial sediment deposited by ice sheets that spread south from Scandinavia during the last **glacial period**.

> **Key term**
>
> **Glacial periods**: historic cold periods associated with the build-up of snow and ice and the growth of ice sheets and glaciers

● Cromer Ridge – a 100-metre ridge just inland from Cromer – is a terminal moraine, a glacial deposition that marks the furthest extent of ice advance.
● Sands and gravels deposited by glacial meltwater streams are found along parts of the coast to the west of the region.
● Till – unsorted glacial sediment – forms thick deposits along much of the coast, spreading some distance inland.

These glacial sediments are extremely weak. They are prone to mass movement and are rapidly eroded by the sea. In parts of the east of the region, cliffs are retreating by over 1 metre each year.

Climate In geological terms, changes in climate have had a significant impact on the North Norfolk coast. During the last glacial period, ice advanced over the area depositing huge thicknesses of sediment. Exposed to geomorphic processes, this has been rapidly eroded in the east and redistributed in the west.

The present-day climate of the region is relatively dry with warm summers and occasionally cold winters. Till dries out and cracks in dry conditions, making it more vulnerable to geomorphic processes. Cold spells in the winter will promote freeze–thaw, particularly if the clay contains deep cracks.

Coastal landforms

The North Norfolk coast exhibits landforms of coastal erosion and coastal deposition. The north-east coast between Overstrand and Happisburgh is dominated by actively eroding till cliffs. In Happisburgh, some people have lost their homes due to rapid coastal retreat. Figure 4 shows the typical profile of these cliffs.

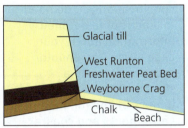

Steep cliffs due to:
- weak rock (till)
- cliffs exposed to powerful waves and long fetch to north-east
- limited beach, leaving cliffs exposed to the full force of waves
- powerful waves undercutting the cliff leading to collapse.

Figure 4 Typical North Norfolk cliffs

The north coast to the west of Sheringham is a coastline dominated by deposition. Longshore drift transports sediment from east to west, forming an extensive spit at Blakeney (Blakeney Point). Sand dunes (for example, at Holkham) and salt marshes (for example, at Stiffkey) are also features of this stretch of coastline.

Figure 5 Blakeney Point spit

Impact of human activity and management

The North Norfolk coast is widely used by people:
- There are many villages and small towns on the coast, linked by main roads. Economic activities include fishing, farming and forestry.
- Much of the area is popular with tourists who visit the coast to enjoy the beautiful landscapes,

taking boat trips and hiking or cycling along the lanes and paths.
- The varied and undamaged coastal habitats, such as the salt marshes and sand dunes, are particularly popular with visitors.

In common with other coastal regions in England and Wales, the North Norfolk coast is managed to balance the social, economic and environmental demands on the coast. Planners face several physical challenges, including high rates of coastal erosion, occasional storm surges (most recently in 2013) and long-term sea-level change associated with **climate change**. A number of coastal defence schemes have been introduced to address the issues:
- At Holkham, the local landowner (the Holkham Estate) has planted pine trees to help stabilise the sand dunes. The Estate has also constructed boardwalks to enable visitors to access the dunes without damaging the vegetation or disturbing the wildlife.
- At Wells-next-the-Sea, groynes (barriers constructed perpendicular to the coast) have been constructed to protect the beach huts; **gabions** (metal cages filled with rocks) have been used to help protect the National Coastwatch Institution lookout station. These forms of hard engineering trap sediment, building up the beach to protect the coastline from powerful waves. By interfering with sediment transfer, such measures can have harmful knock-on effects further along the coast where beaches can become starved of sediment.
- Several hard engineering measures have been adopted in Cromer, Sheringham and Overstrand to protect the coastline, including **sea walls** and rock armour (piles of massive boulders). Figure 6 shows the coastal defences at Overstrand.

Key terms

Climate change: changes in long-term temperature and precipitation patterns that can either be natural or linked to human activities

Gabion: metal cages filled with rocks, which can form part of a sea defence structure or be placed along rivers to protect banks from erosion; an example of hard engineering

Sea walls: curved concrete structures placed along a sea front, often in urban areas such as the front of a promenade, designed to reflect back wave energy; an example of hard engineering

Figure 6 Coastal defences at Overstrand

Coastal defences – particularly hard engineering schemes – have an impact on natural processes. Groynes interrupt the movement of sediment by longshore drift, and sea walls can deflect high-energy waves along the coast. It is often the case than coastal defences protect one area but cause increased problems elsewhere along the coast. Some people consider these artificial structures to be ugly, ruining the natural landscape.

Revision activity

Draw a simple outline map of the North Norfolk coast and use appropriately located labels to describe the geomorphic processes operating on the coast.

Exam tip

For this part of the specification you do need to learn some detailed information about your case study. You should practice drawing a sketch of the coastline (Figure 3), locating the main settlements and coastal landforms.

Now test yourself

1 Outline the impacts of geology on the geomorphic processes operating on the North Norfolk coast.
2 Use Figure 4 to identify the factors responsible for shaping these cliffs on the North Norfolk coast.

Answers online

TESTED ☐

Exam practice

1 With reference to your chosen coastal case study, describe how geomorphic processes are affected by geology and climate. [6]
2 Use Figure 6 to describe the coastal defences at Overstrand. [4]
3 With reference to Figure 6 and your own knowledge about your chosen coastal case study, explain how coastal management works with geomorphic processes to impact the landscape. [6]

ONLINE ☐

2 People of the UK

The UK's major trading partners

The UK's major trading partners

What is trade?

Trade involves the movement of goods and **services** across the world. Trade involves **imports** (products brought into a country) and **exports** (products taken out of a country). Typically trade involves transport by plane, container ships or trains. Increasingly modern trade involves the internet, for example in finance and the creative industries.

The UK has a long tradition of trading with other countries.
- As an island, much of the UK's trade involved ships and many coastal settlements developed into thriving ports.
- In the past, the UK traded a great deal with its colonies across the world, importing cotton, grain and products from the tropics such as fruit, coffee and spices.
- In return, the UK exported mainly manufactured products.

> **Key terms**
>
> **Trade**: the buying and selling of goods and services between countries
> **Services**: a function, or 'job', that an ecosystem provides
> **Imports**: the purchase of goods from another country
> **Exports**: the selling of goods to another country

Who are the UK's trading partners?

Today the UK's most important trading partners are members of the European Union (EU). As a single market, goods can be traded without tariffs between member states. The USA is an important historic trading partner. In recent years, there has been a significant growth in trade with China. Trade is also important with former colonies, now members of the Commonwealth.

It is important to try to have a balance between exports and imports. If imports exceed exports, a country has a **trade deficit**. In the long-term this can be expensive for the country. The UK has a trade deficit of about £2.0 billion per year. Recently, this has fallen slightly due to a reduction in imports and an increase in the export of manufacturing products.

> **Key term**
>
> **Trade deficit**: the amount by which a country's imports exceed the value of its exports

Imports to the UK

Look at Figure 1. It shows where imports to the UK come from. Notice the dominance of the EU – in particular Germany – but also the importance of the USA and China. Post–Brexit (2019) it will be important for the UK to forge trading links with a large number of countries beyond the EU.

Figure 1 Top ten trading partners for imports to the UK (2014)

Country	Imports (£bn 2014)
Germany	56
Netherlands	34
USA	32.1
China	31.6
France	24.8
Belgium	20.8
Irish Republic	18.2
Norway	16.8
Italy	15.2
Spain	12.5

- The bulk of the UK's imports by value are manufactured products, in particular cars, electrical items and clothing.
- Most of the items imported from China – clothing and electronics – are cheaper than alternatives made in the UK.

- The UK is a relatively wealthy economy so expensive items such as cars from Germany have a large market.
- Petroleum and petroleum products are important imports, providing the UK with fuel and a versatile raw material for the chemicals industry.

Exports from the UK

Whilst the USA is the main export destination, most of the UK's exports go to Europe and, in particular, to EU countries (Figure 2). Exports to China are increasing and, just outside the top ten, so are exports to the United Arab Emirates (UAE).

The UK's main exports are engineering, oil and vehicles. The fastest growing exports in 2014 were gems, precious metals and coins. The UK's manufacturing sector – motor vehicles and aeronautical engineering – also showed an increase in exports in 2014.

Figure 2 Top ten UK export countries (2014)

Country	% of total UK exports by value
United States	11.8%
Germany	9.8%
Netherlands	7.4%
Switzerland	6.6%
France	5.9%
Ireland	5.9%
China	5.1%
Belgium	4.1%
Italy	2.8%
Spain	2.8%

Exam practice

1 Calculate the value of the USA's imports to the UK as a percentage of that imported from Germany. [1]
2 Use Figure 2 to describe the pattern of exports from the UK. [4]
3 To what extent are healthy trading links important for economic growth in the UK? [6]

ONLINE

Now test yourself

1 Define the following terms: trade, imports, exports and trade deficit.
2 Germany is the UK's most important trading partner. True or false?
3 Why has trade with China increased in recent years?
4 Why is petroleum an important import into the UK?

Answers online

TESTED

Revision activity

Use the information in Figures 1 and 2 to draw simple flow maps to show imports to and exports from the UK.

Exam tip

Remember that when you are asked 'to what extent' in a question, you must make a judgement and then be prepared to justify it. 'To what extent' requires you to decide where you are on an imaginary line from 0% agree (100% disagree) to 100% agree.

Diversity in the UK

What makes the UK special?

The UK is a country of immigrants. Through the ages people from all over the world have arrived in the UK bringing with them a rich cultural heritage. This continues today as the UK welcomes people from war-torn parts of the world, such as Syria, and from other countries in the EU. This accounts for the UK's rich heritage and its cultural, social and economic diversity.

The UK is a country of physical and human contrasts. There are some significant variations across the UK at both regional and local scale.

Employment patterns

Over the last 25 years, the patterns of employment have changed a great deal. The main trends are shown below:

1 Women are encouraged to work and follow careers

- Employers offer flexible working hours and often help with child care.
- The government provides financial support for child care to enable women to return to work.
- Work places increasingly provide crèche and child care facilities on site.

3 Flexible working hours

- People increasingly choose to operate flexible working hours, maybe combining office work with working from home.
- Some people work during the evenings or overnight.
- The availability of mobile phones enables people to work on the move.

2 Increasing numbers of people are part time or self-employed

- People seeking a better work-life balance prefer to work from home.
- The widespread availability of high-speed internet access enables people to work on the move (using tablets or mobile phones) or from home.

4 De-industrialisation

- As the UK has de-industrialised, jobs in manufacturing have been replaced by employment in the services or tertiary sector, such as education and health care.
- A new quaternary (knowledge-based) sector has developed with jobs in research, information technology and the media.

Figures 1 and 2 show the patterns of employment in manufacturing and services. Notice the following:

- The importance of the service sector compared with manufacturing
- Manufacturing is largely concentrated in the traditional industrial heartlands of the Midlands and northern England
- Services are concentrated in London and the southeast of England, urban centres and centres of tourism (for example, parts of Wales).

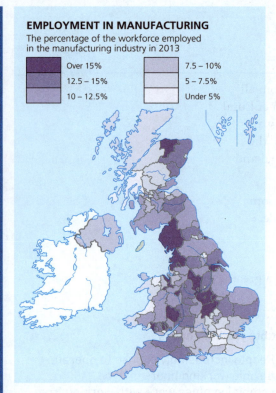

EMPLOYMENT IN MANUFACTURING
The percentage of the workforce employed in the manufacturing industry in 2013

- Over 15%
- 12.5 – 15%
- 10 – 12.5%
- 7.5 – 10%
- 5 – 7.5%
- Under 5%

EMPLOYMENT IN SERVICES
The percentage of the workforce employed in the service industry in 2013

- Over 90%
- 85 – 90%
- 80 – 85%
- 75 – 80%
- Under 75%

Figure 1 Employment in manufacturing, 2013   **Figure 2** Employment in services, 2013

Patterns of average income

Whilst the UK's average salary is about £26,500, there are huge disparities in incomes in the UK.

- Top Premier League footballers now earn over £500,000 a week.
- The chief executives of the top 100 FTSE companies earn the equivalent of the UK's average yearly salary in just two and a half days.
- 80% of new jobs have salaries of less than £16,640 per year for a 40-hour week. (A high proportion of these are working in vital health, education and social services.)
- Working on the minimum wage of £6.70/hour brings an annual salary of just over £13,000.

Disposable income

Disposable income is the money people have to live on once their taxes, pensions and mortgage/rent have been paid. The UK's average disposable income is about £17,500. There are huge variations across the UK and within regions.

- The greatest range is in London (£16,801 in Barking and Dagenham compared with £43,577 in Westminster).
- Residents in Westminster have, on average, four times the disposable income of residents in Leicester.
- The Midlands and Northern Ireland have the lowest disposable incomes in the UK.

Now test yourself

1 State one reason for more women being in work.
2 State one reason why there has been an increase in self-employment.
3 How has mobile-phone technology led to flexible working?
4 Using Figure 2, give three locations where over 90% of the workforce is employed in services.

Answers online

TESTED

Life expectancy

The average **life expectancy** in the UK is now 81 years. It has risen consistently in recent years as health care, diets and standards of living have improved. People are now living more than 5 years longer, on average, than in 1990.

However, life expectancy is not equal across the UK (Figure 3).
- It is highest in the South East (82.4 years) and lowest in Scotland (79.1 years).
- Men living in Blackpool live an average of 8 years fewer than men living in the City of London and 9 years fewer than men living in South Cambridgeshire, which has the highest life expectancy.
- In England, women living in Manchester have the lowest life expectancy.

These disparities reflect variations in incomes and quality of life across the UK. Poor diets and smoking are the biggest risks leading to premature death or disability.

Now test yourself

TESTED ☐

1 Study Figure 3.
  (a) Which region has the lowest life expectancy?
  (b) Does the East Midlands or West Midlands have the longer life expectancy?
  (c) Which region has the highest life expectancy?
  (d) Calculate the difference in life expectancy between the highest and the lowest region.
2 How do lifestyle choices affect life expectancy?

Answers online

Educational attainment

In 2015 across England, 68.8% of GCSE entries gained an A★ to C grade. This was a slight increase on the previous year. However, there were wide regional variations.
- London recorded the best results, with 72% achieving A★ to C grades.
- The highest values were in the rich London boroughs, such as Kensington and Chelsea (73.1%).
- The region of Yorkshire and the Humber recorded just 65% A★ to C grades.
- The lowest values were in towns in northern England. such as Knowsley (40.8%) and Middlesbrough (40.9%).

There is a clear link between poverty and educational achievement. The most deprived areas, with low incomes and high unemployment, tend to have to lowest levels of achievement.

Ethnicity

Ethnicity is about groups of people who share common roots, often based on culture, religion or nationality. In the UK, ethnic groups tend to be immigrants associated with foreign nationalities such as Bangladeshi or Pakistani.

Whilst ethnic groups have settled widely across the UK, there is a concentration in major cities and, in particular, in London where there are about 3 million foreign-born residents (Figure 4).

Key term

Life expectancy: the average number of years a person might be expected to live

Figure 3 Life expectancy by UK region, 2015

Region	Average life expectancy (years)
South East England	82.4
East of England	82.2
South West England	82.0
Greater London	81.4
East Midlands	81.2
West Midlands	80.9
Yorkshire and the Humber	80.6
Wales	80.3
North East England	80.1
North West England	80.0
Northern Ireland	79.6
Scotland	79.1

Key term

Ethnicity: relates to a group of people who have a common national or cultural tradition

Increasingly, residential areas in the UK are becoming ethnically mixed. However some ethnic groups do tend to form distinct clusters in cities, especially Pakistanis, Bangladeshis and some Black Africans. With low incomes and limited job security ethnic groups may find themselves living in relatively deprived areas in the inner city.

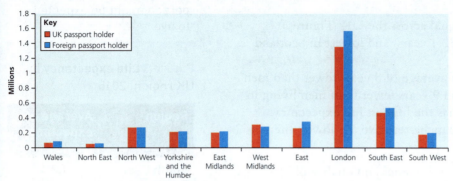

Figure 4 Foreign-born residents in England and Wales, 2011

Study Figure 4.

(a) Which two regions show the lowest number of foreign-born residents?

(b) What is unusual about the numbers of foreign-born residents in the West Midlands compared with other regions in England and Wales?

(c) Approximately how many UK-passport-holder foreign-born residents are there in the South East?

Answers online

Access to broadband

Today almost 100% of households in the UK can access the internet and over 90% of households have access to superfast (30Mbps+) broadband.

The pattern of broadband availability in the UK broadly reflects the pattern of the UK's population.

- High availability in London and the South East, together with the cities in the north of England
- Low availability in the more remote parts of the UK (for example Wales, Scotland and South West England)
- Numerous small pockets of low availability across the UK, including within those areas with a generally good availability (for example parts of London and the South East).

Causes and consequences of development within the UK

What is the pattern of development in the UK?

REVISED ☐

Patterns of household wealth

Development in the UK is not even. Figure 1 shows the so-called **North–South divide** in the UK as measured by household wealth.

- There is a clear wealth divide between the wealthy 'south' and poorer 'north'.
- Disparity exists within the 'south' – the South-East region is much wealthier than the other regions – over 13 per cent of households have a total wealth of close to £1m a year!

> **Key term**
>
> **Development**: the state of growth or advancement whereby people and places improve over time

Patterns of health

Figure 2 shows the deaths resulting from coronary heart disease. This health indicator also shows a clear North–South divide. High rates of coronary heart disease are most often associated with lifestyle issues including smoking, diet and exercise. Poverty and standards of living are also important factors affecting people's health.

Imbalances within regions

Despite the clear regional trends indicated by Figure 1 and 2, regional averages do hide huge imbalances. Within all regions there will be some very wealthy individuals and some who are very poor. For example, in London:

- There are over 7500 people who sleep on the streets each night.
- The borough of Tower Hamlets is the third most deprived borough in England and has the highest rate of child poverty in the country.
- In the borough of Islington, the mortality rate for coronary heart disease is 114 per 100,000, higher that any of the regional figures in Figure 2.

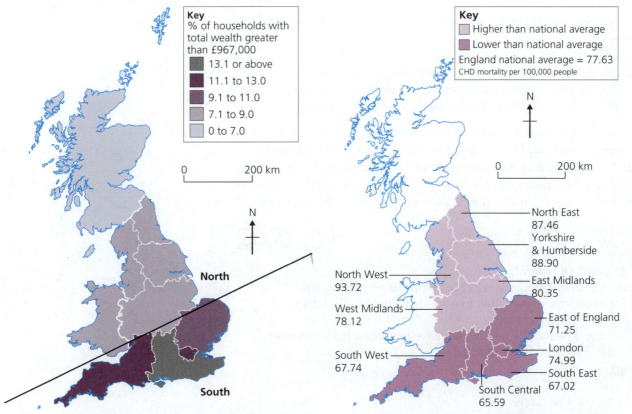

Figure 1 The North-South divide (household wealth)   **Figure 2** Patterns of coronary heart disease

Now test yourself

Study Figure 1 and 2. Consider whether the following statements are true or false.

(a) London has the highest percentage of households with total wealth greater than £967,000.

(b) In Wales, 7.1–9.0 per cent of households have total wealth greater than £967,000.

(c) The North East has the highest rate of coronary heart disease.

(d) The rate for coronary heart disease in the South West is higher than in the South East but lower than in London.

(e) The rate of coronary heart disease for London is 3.64 per cent higher than for the East of England.

Answers online

What are the causes of uneven development?

REVISED

Geographical location

London is the centre of economic activity and wealth creation in the UK, mainly due to its role as capital of the UK.

- London has long been one of the world's major trading centres.
- It is the UK's major hub for business, finance and media.
- Many national and international companies have their headquarters in London.

As London has grown, wealth has extended out into the rest of the South East.

- Many people who work in London now commute from the 'home counties', choosing to live in more pleasant **rural** surroundings.
- The areas around London have witnessed tremendous economic growth.
- Cities like Cambridge have become core growth centres.

Increasingly, London and the South East has benefited from trade with Europe.

- London and the South East have excellent access to the continent.
- There are fast Eurostar rail services through the Channel Tunnel.
- There are several ferry routes and many air connections from London's airports, including City Airport in the centre of London.

Regions in the north and west of the UK are more distant from the European mainland. Despite good transport links with cities such as Manchester and Glasgow, many rural areas are remote and inaccessible. Whilst some northern cities have a wealthy base, the rural districts are often much poorer, unlike the wealthy rural areas in the South East.

> **Key term**
>
> **Rural**: areas that are not urban; characteristic of the countryside rather than towns and cities

Now test yourself

TESTED

1 Give one reason why:
   (a) London is the centre of economic activity in the UK
   (b) economic growth has extended beyond London in to the surrounding regions
   (c) London and the South East have benefited from proximity (closeness) to Europe.

2 Identify **four** facts about London's Crossrail project.

Answers online

Economic change

Time period	Economic changes
Before Industrial Revolution	Most people in the UK worked in farming, mining or related activities – the primary sector.
Industrial Revolution (late 19th century)	People moved to the towns and cities for work – making steel, ships or textiles (the manufacturing sector). During this period, much of the UK's growth was centred on the northern coalfields. Heavy industries and engineering thrived in the cities and a great deal of wealth was generated.
De-industrialisation (mid-late 20th century)	Many industries in the 'north' closed, mainly due to competition from abroad, and people lost their jobs. In 2015, the Redcar steelworks on Teesside closed with the loss of some 1700 jobs. This had a huge impact on the community and the economy of the region.
Late 20th century-early 21st century	Recently there has been huge growth in the service or **tertiary sector** involving jobs in health care, offices, financial services and retailing.
	Most recently, the **quaternary sector** has developed with jobs in research, information technology and the media. Most of these jobs have been based in London and the South East. Today London is a world centre for financial services, media, research and the creative industries, and it has benefited hugely from **globalisation** and interconnectivity with the rest of the world.

Key terms

Tertiary industries/sector: service industries and jobs such as teaching; very few people are employed in this sector in a developing country

Quaternary industry: work in the 'knowledge economy' that involves providing information and the development of new ideas

Globalisation: the process whereby places become interconnected by trade and culture

Revision activity

Use the information in the table above to sketch a timeline describing the main economic changes in the UK. This could be illustrated with sketches or photos to help you remember the key points.

Infrastructure

Infrastructure involves transport, services and communications. In recent years, London and the South East has benefited from a number of developments including:

- the Channel Tunnel (1994)
- expansion of airports, such as Stansted, and the construction of new terminals, such as Terminal 5 at Heathrow (2008)
- High Speed 1 Eurostar trains starting to operate from London St Pancras (2007).

In the future, there are several planned developments in transport, including Crossrail and the construction of a new airport runway, probably at Heathrow.

Key term

Infrastructure: the basic structures and facilities needed for a society to function, such as buildings, roads and power supplies

Crossrail

- Crossrail is one of Europe's largest construction projects and is due to be completed in late 2018.
- Costing some £15 billion, the new 100 km rail route will run from Reading and Heathrow in the west to Shenfield and Abbey Wood in the east, passing through Central London in a series of tunnels.
- Linking London's key employment, leisure and entertainment districts, it will carry some 200 million passengers a year and will add an estimated £42 billion to the economy of the UK.
- Crossrail will support regeneration projects and cut journey times across the capital.

Government policy

Many companies – both UK based and international – have chosen to be in London rather than elsewhere in the UK. Government investment in infrastructure projects such as Crossrail, the regeneration of London's docklands (1980s+) and construction of the Olympic Site (2012) have all promoted the economic growth of the 'south'.

In 2015, the government announced plans to create a Northern Powerhouse of modern manufacturing industries specialising in science and technology across the major cities of the north. The aim is to redress the North–South economic imbalance, and to attract investment into northern cities and towns. Several transport improvements will support this initiative:

- HS2 (High Speed 2) is a £50 billion project to build a new high-speed railway line to connect London with Birmingham and then on to Sheffield, Leeds and Manchester. It may then be extended to Newcastle and into Scotland. The scheme, which is due to start in 2017 for completion in 2033, is highly controversial as the route passes through several stretches of highly valued countryside and close to many people's homes.
- Electrification of the Trans-Pennine Express Railway between Manchester and York by 2020, reducing journey times by up to 15 minutes and completing the electrified link between Liverpool and Newcastle.
- Electrification of the Midland Mainline, between London and Sheffield by 2023.

Now test yourself

TESTED

1 What is the 'Northern Powerhouse'?
2 Give **one** example of government policy that has promoted economic growth in London.
3 Outline **one** transport project that is intended to boost economic development in the north of the England.

Answers online

Exam practice

1 (a) Outline the meaning of the term 'north-south divide'. [2]
  (b) Suggest how patterns of wealth and health can illustrate regional inequality in the UK. [4]
2 Suggest reasons for the economic dominance of London and the South East. [6]
3 To what extent is government policy addressing the issue of the 'north-south divide' in the UK? [6]

ONLINE

Exam tip

Remember that when you are asked 'to what extent' in a question, you must make a judgement and then be prepared to justify it. 'To what extent' requires you to decide where you are on an imaginary line from 0% agree (100% disagree) to 100% agree.

Case study: Economic growth and decline

In this section you need to study **one** place or region in the UK which has experienced economic growth and/or decline.

Case study: Salford Quays, Manchester

Where is Salford Quays?

Salford is located to the west of the city of Manchester, in the heart of northern England. It is an inner-city urban environment and is home to some 200,000 people.

Figure 3 Location of Salford Quays and Manchester

Early economic growth

● During the industrial revolution, Manchester became a centre for the processing of cotton, imported from the USA. It developed thriving manufacturing and engineering sectors, specialising in the production of machines for factories.

● In 1894, the Manchester Ship Canal opened, linking the city centre with the River Mersey and the Irish Sea. At the head of the Manchester Ship Canal a massive 90-hectare complex of inland docks (Salford Quays) was constructed to accommodate the thriving trade.

● Thousands of people were employed in the docks and a large community became established with homes, factories and shops.

Mid-twentieth-century decline

● In the 1960s and 1970s, the new larger container ships were unable to use the Manchester Ship Canal, plunging the area into decline.

● Over 3000 people lost their jobs and the docks finally closed in 1982.

● The derelict land was heavily contaminated, houses had fallen into disrepair and there were high rates of unemployment and crime.

Recent redevelopment and growth

● In the mid-1980s, the newly named Salford Quays received funding from the UK's Urban Programme to reclaim the area for commercial office and residential use. In 1985, the Salford Quays Development Plan was launched. Since then, massive investment has resulted in new homes, education and health facilities, new businesses and shops. There are city parks, cleansed waterways and green spaces.

● One of the most ambitious and successful projects is the Lowry Building. Completed in 2000 at a cost of £64 million, the building houses an 1800-seat theatre and several galleries, bars and cafés. It provides a permanent home for the paintings of local artist JS Lowry, after whom the building is named.

● In 2007, the BBC moved five of its departments, including BBC Sport and BBC Radio 5 Live, to a new development on Pier 9 called MediaCityUK. Constructed at a cost of £550 million, MediaCityUK created 10,000 jobs and added an estimated £1bn to the regional economy by 2013.

● Other developments in Salford Quays include a £90m retail and leisure facility called the Lowry Outlet, the construction of the Imperial War Museum of the North and the construction of a new line of the city's Light Rapid Transit (LRT) tram system.

Salford City Council aims to establish Salford as a 'Modern Global City' by 2025, with Salford Quays being very much as the heart of its regeneration plans.

Figure 4 Aerial view of the BBC's MediaCityUK

Now test yourself

TESTED

1 (a) In which city is Salford Quays located?
  (b) What canal linked Salford Quays to the Mersey and the Irish Sea?
  (c) Why did the docks decline in the mid-twentieth century?
2 (a) What is the Lowry Building?
  (b) Name three other developments at Salford Quays.
3 Why was it important to link Salford Quays to the Manchester LRT tram system?

Answers online

Exam practice

1 Using a case study, describe the consequences of economic growth.
[6]

ONLINE

The UK's changing population

Changes in the UK's population structure

Measuring population – the census

The population of a country is measured by a survey called a **census**. In most countries of the world, including the UK, a census is carried out every ten years. The most recent UK census took place on 27 March 2011.

The UK census provides a valuable snapshot of the population and its characteristics, helping the government to plan funding and public services for the future. Figure 1 shows the population of the UK by country and Figure 2 shows the growth of the UK's population from 1901 and its projected population for 2021.

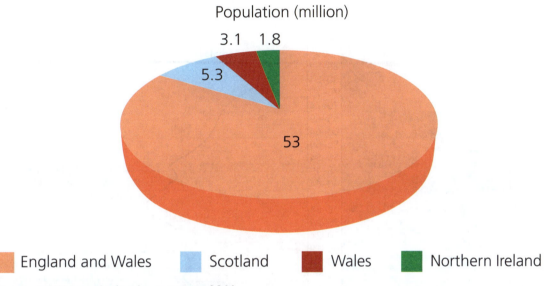

Population (million)

England and Wales Scotland Wales Northern Ireland

Figure 1 UK population by country, 2011

Year	UK population (millions)
1901	38.2
1911	42.1
1921	44.0
1931	46.0
1951	50.2
1961	52.7
1971	55.9
1981	56.4
1991	57.4
2001	59.1
2011	63.2
2021	67.6 (projected)

Figure 2 Population growth of the UK, 1901–2021

Now test yourself

In 2011, the total population of the UK was 63.2 million. Calculate the number of people living in England and Wales as a percentage of the UK's total population.

Answers online

TESTED

The UK's population structure

The breakdown of a population by age and sex is called the **population structure**. It is commonly illustrated by a diagram called a **population pyramid**.

- Bars are drawn to represent each five-year age band.
- The length of each bar relates to the number of people of that age in the population.
- Bars are drawn for both males and females.

Figure 3 shows the most recent population pyramid for the UK based on the 2011 census. There is also an outline for 2001 to show how it has changed. Population pyramids can be used to see trends in the population, such as declining birth rates or increases in the number of elderly people. These trends provide useful information for the government in helping to plan for future education, housing, employment and health care needs.

Key terms

Population structure: the composition of a population

Population pyramid: a diagram, essentially a bar graph, that shows the structure of a population by sex and age category that may resemble a pyramid shape

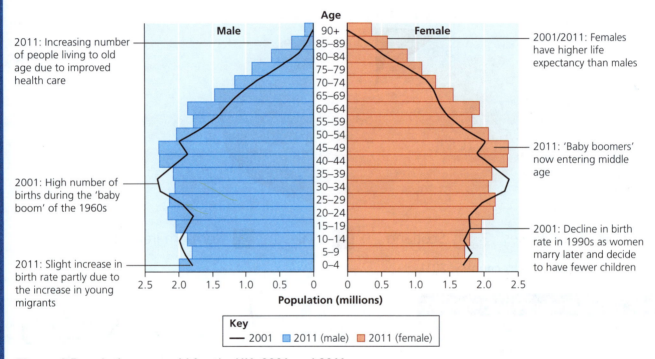

2011: Increasing number of people living to old age due to improved health care

2001/2011: Females have higher life expectancy than males

2001: High number of births during the 'baby boom' of the 1960s

2011: 'Baby boomers' now entering middle age

2011: Slight increase in birth rate partly due to the increase in young migrants

2001: Decline in birth rate in 1990s as women marry later and decide to have fewer children

Key
— 2001 ■ 2011 (male) ■ 2011 (female)

Figure 3 Population pyramid for the UK, 2001 and 2011

Now test yourself

TESTED ☐

Study Figure 3. List three changes in the UKs population pyramid between 2001 and 2011.

Answers online

Changes in the UK's population structure since 1900

Since 1900, the main trends in population have been:
- Decrease in the proportion of young people (aged 0–14). Birth rate has decreased as infant mortality rates have declined and increasing numbers of women have chosen to follow careers.
- Increase in the elderly population, with over 16 per cent of the UK's population now aged 65 and over. This proportion is expected to increase further. Improved health care, higher standards of living and better lifestyles (especially non-smoking) have contributed to this trend.

The divided bar graph (Figure 4) illustrates these changes.

Date	Population details
Early 1900s	In 1990, the population of the UK was about 42 million and 30% of people lived in the countryside, mostly engaged with farming.
	Farming was becoming **mechanised** – increasing numbers of people were moving to work in factories in towns and cities.
	In 1911, about 30% of the UK's population was aged 0–14 years. This reflected a relatively high birth rate. Many children worked on farms. Few women had careers at the time and contraception was not widely used.
	Relatively few people lived to old age due mainly to limited health care.
Mid-1900s	The youthful population 'bulge' in 1921 moved up the pyramid, increasing the proportion in middle age.
	More people were living into old age.
	After a period of declining births, there was a sudden increase – this continued into the early 1960s and became known as the 'baby-boomer generation'.
Late 1900s	By the late 1990s, the 'baby-boomers' were moving into middle age.
	Births were generally low and steady because women were following careers and choosing to have smaller families.
	The number of older people rose – increase reflecting higher standards of living and high quality medical care.
Early 2000s	Birth rate remains low and steady.
	The 'baby-boomers' have moved into late middle age.
	The number of older people remains high and is growing – average life expectancy is now well into the 80s.

Key term

Mechanisation: the process whereby machinery is introduced to complete work normally done by hand, e.g. washing machines, tractors, industrial robotics, engines, automated tools, etc.

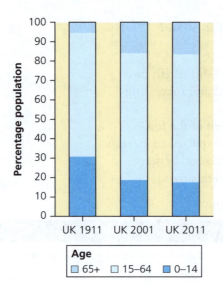

Figure 4 Divided bar graph showing the UKs population structure, 1911–2011

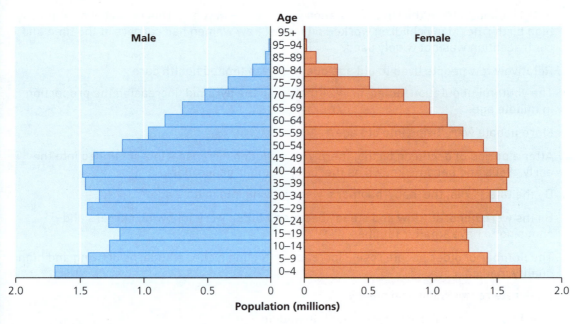

Figure 5 UK's population pyramid, 1951

Now test yourself

Give reasons for the two main trends in the UK's population structure from 1900:

(a) decrease in the proportion of young people (aged 0–14)

(b) increase in the proportion of older people (aged 65+).

Answers online

Revision activity

Construct a diagram, such as a timeline, to outline the main changes in the UK's population structure from 1900.

The Demographic Transition Model

The **Demographic Transition Model** is a graph showing the typical changes that take place in a country's population over time. Figure 6 shows the changes that have taken place in the UK since 1700. Notice that there are three lines on the graph:

- **Birth rate** – the number of live births per 1000 of the population per year.
- **Death rate** – the number of deaths per 1000 of the population per year.
- **Total population** – the total population of the country.

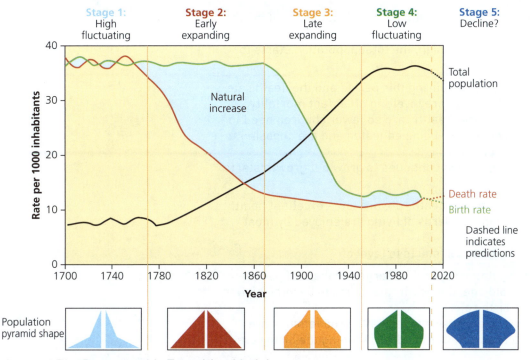

Figure 6 The Demographic Transition Model

The difference between the birth rate and the death rate is called the **natural increase**. This is usually expressed as a percentage. Notice that the natural increase is shaded on the graph between the birth rate line and the death rate line. The total population of a country is the natural increase +/− **migration**. In the UK, migration has contributed quite a lot to the growth of the population.

Key terms

Demographic transition model: a theoretical model based on the experience in the UK showing changes in population characteristics over time more arid and desert-like, usually because of drought, deforestation, over-cultivation or over-extraction of water
Natural increase/decrease: the difference between the birth rate and death rate, usually expressed as a percentage
Migration: the movement of people from one place to another; may be voluntary or forced, permanent or temporary, domestic or international

The Demographic Transition Model can be divided into a number of stages.

Stage	Population changes
1	• Both birth rate and death rate are high. Disease and poor living standards result in a high death rate. Children support the family and, because many die in infancy, lots are born to guarantee that a few will survive. • Birth and death rates cancel each other out so the total population remains stable.
2	• Death rate drops as health care and standards of living improve. • As birth rate is still high, the total population starts to increase.
3	• Birth rate declines during this period and the death rate continues to fall before levelling off. Infant mortality falls due to better health care so fewer children need to be born. Women are being educated and are choosing to have fewer children. • The total population continues to grow but growth starts to slow down.
4	• Both birth rate and death rate are low - excellent health care and high standards of living are enjoyed by most people. • The total population starts to level off.
5	• As the population ages, with large numbers of people reaching old age, the death rate starts to become higher than the birth rate. • Children are expensive to look after and families are deciding to have fewer children, keeping the birth rate low.

The UK is currently in Stage 4 of the Demographic Transition Model. In recent years, immigration has led to a slight increase in the country's birth rate and to its total population. The large number of elderly people means that death rate will probably increase slightly.

Now test yourself

TESTED ☐

Study Figure 6, which shows the Demographic Transition Model.
(a) Describe the birth rate and death rate in Stage 1.
(b) Why does the total population increase in Stage 2?
(c) What date did Stage 2 end and Stage 3 begin?
(d) Why does the total population start to level off in Stage 4?
(e) What happens in Stage 5?

Answers online

Ageing population in the UK

In 2011, 9.2 million people (16 per cent of the UK's population) were aged 65+. This is almost a million more people than in 2001. As the bulge in population – the 'baby boomers' – born in the mid-1900s move into old age, the UK is set to experience an **ageing population**.

An ageing population presents challenges and opportunities.

> **Key term**
>
> **Ageing population**: population structure that becomes distorted with a high and increasing proportion of people in middle and old age

Challenges	Opportunities
● Elderly people have greater medical needs and the costs of looking after them will increase in the future. ● They will need increasing amounts of care to enable them to stay in their own homes. ● Their children – in middle age – will increasingly be responsible for their care.	● Many older people give up their time to work as volunteers in the community and some continue to work in paid employment. ● Many newly retired people enjoy good health and have money to spend on travel, home improvements and hobbies such as gardening. ● Businesses specialise in providing services for older people.

> **Revision activity**
>
> Draw a spider diagram to summarise the causes, effects and responses to an ageing population in the UK.

The table below summarises the causes, effects and responses of the UK's ageing population.

Causes	Effects	Responses
● A large number of people born after the Second World War and through into the 1960s ('baby boomers') are now moving into old age ● Improved health care and new treatments prolong life especially from diseases such as cancer and heart conditions ● Reductions in smoking, which caused a huge early death toll in the past ● Greater awareness of the benefits of a good diet and regular exercise ● Many older people are reasonably well off financially so can afford a reasonable standard of life	● Health care costs are very high ● Shortages of places in care homes many of which are expensive ● Many older people are looked after by their middle-aged children, often affecting their ability to remain in full-time employment ● Older people are valued employees as they have high standards and are reliable ● Older people act as volunteers in hospitals, advice centres and food banks ● Many older people are keen to travel and to join clubs, societies, sports centres, etc. This helps to boost the economy and provide jobs	● Government issued Pensioner Bonds in 2015 to encourage older people to save money for the future ● Pensioners receive support in the form of care, reduced transport costs and heating allowances (winter fuel payments) ● Retirement age, which used to be 65, is being phased out to encourage people to continue working ● State pension age is gradually being increased to 67 and will probably rise further ● Pronatalist policies introduced to encourage an increase in birth rate to balance the population structure. This could include cheaper child care, improved maternity and paternity leave and higher child-benefit payments

Immigration into the UK

Immigrants are people who migrate into a country, whereas **emigrants** are people who move out of a country. The UK is known for its fair and welcoming attitude towards people from all over the world. This explains why the UK has such a diverse cultural heritage.

In the twentieth century, the UK welcomed people from the Caribbean and from India, Pakistan and Bangladesh. In the twenty-first century, the UK has welcomed people migrating from other parts of Europe, Asia and from war-torn countries such as Syria.

> **Key terms**
>
> **Immigrants:** people who move from one country to settle in another
> **Emigrants:** people who leave one country to settle in another

What are the recent trends in immigration?

In the year to March 2015, net migration (the difference between immigration and emigration) reached 330,000, an all-time high. Much of this was due to people moving into the UK from poorer parts of Europe (such as Poland and Lithuania) and from conflict areas such as Afghanistan, Iraq and Syria.

The majority of people moving to the UK are seeking employment – they are often able to earn more in low-income jobs in the UK than in their home country. Others travel to the UK as students or to join family members.

Where do people come from?

In the 2011 census, the top country of origin for migrants to the UK was India (about 700,000), followed by Poland (580,000) and Pakistan (480,000).

- Many immigrants from India and Pakistan travelled to join family who were already living in the UK.
- Recently, a large number of people from Poland have decided to move to the UK in search of better wages and improved opportunities.
- A considerable number of people have come to the UK from other European countries. (This freedom of movement may change when the UK leaves the EU in 2019.)
- Migrants have arrived from across the world, including Africa, Asia, the USA and the Caribbean. Several countries are former colonial countries and current members of the Commonwealth (for example Jamaica).

Trends since 2011 include:

- In the year ending March 2015, the highest numbers of migrants from outside the EU were from China.
- From within the EU, large numbers migrated from Romania and Bulgaria, two of the poorest and most recent countries to join the EU.
- In the last few years, an increasing number of people have arrived from war-torn countries, such as Syria and Afghanistan. They are seeking asylum (safety), fearing for their lives if they return to their countries of origin.
- In the year to June 2015, the UK received over 25,000 asylum applications, an increase of 10 per cent on the previous year.

Now test yourself

1 What is the difference between immigration and emigration?
2 Suggest why people have migrated to the UK from:
   (a) India and Pakistan
   (b) Romania and Bulgaria
   (c) Syria.

Answers online

What are the social and economic impacts of immigration on the UK?

	Advantages	Disadvantages
Social	• Introduction of different cultures including foods, music and fashion • Immigrants bring skills that may be in short supply within the UK • Immigrants are often keen to engage with the local communities	• May be some tensions with local people or other ethnic groups • May be some bad feeling about housing shortages leading to social unrest • Some people feel that the UK is already overcrowded and that too many immigrants will lead to increased urban pollution and congestion
Economic	• Workers pay taxes to the government – the majority of immigrants work – more money is paid in taxes than received in benefits • Immigrants often take low-paid jobs in farming, factories or support services such as cleaning. Semi-skilled workers have filled gaps in the building industry as well as working as nurses • Some immigrants are well educated and highly trained • Those immigrants who study in the UK pay a considerable amount to colleges and universities	• Extra costs for health care, education and social services • House prices and rents may increase as demand outstrips supply • Money may be sent home by immigrants so does not get spent in the UK • Some people think that migrants are 'taking our jobs' and increasing unemployment - there is, however, no real evidence that immigration is linked to unemployment

Exam practice

1 (a) What is a population pyramid? [2]
  (b) Explain the changes in the UK's population structure since 1900. [6]
2 Study Figure 6 showing the Demographic Transition Model.
  (a) What is the meaning of the term 'natural increase'? [2]
  (b) Describe the changes in natural increase from 1700–1940. [4]
  (c) Explain why some countries have passed from Stage 4 to Stage 5. [6]
3 (a) Outline the causes of the UK's ageing population. [4]
  (b) Evaluate the options available to the government in responding to the UK's ageing population. [6]
4 Describe the recent trends in immigration into the UK. [4]
5 'Immigration to the UK brings more problems than benefits.' To what extent do you agree with this statement? [6]

ONLINE

Revision activity

Draw a summary diagram to describe the social and economic advantages and disadvantages of immigration to the UK.

Exam tip

Remember that when you are asked to 'evaluate' you need to consider both sides of the argument. Try to consider the advantages and disadvantages and ensure that your answer is balanced.

Causes and consequences of urban trends in the UK

Suburbanisation

Figure 1 Causes and consequences of suburbanisation

Causes	Consequences
• **Suburbanisation** started in the mid- to late twentieth century, when public transport and private car ownership meant that commuters could live further out from the city centre. • There was a move towards home ownership in the UK during the 1970s, which led to private housing estates being built on the edges of cities. Building in these areas allowed people to have more land for gardens and more public open space, compared with housing areas nearer the town centre.	• Local shopping centres have been constructed, along with many primary schools and a smaller number of secondary schools. • Created a demand for out-of-town retail parks. • Buildings in the city are left vacant. These buildings might quickly start to look derelict and be vulnerable to vandalism, preventing inward investment in the city. • Increased congestion and pollution from commuting.

Key terms

Suburbanisation: a change in the nature of rural areas such that they start to resemble the suburbs

Counter-urbanisation: the movement of people from urban areas into rural areas; these may be people who originally made the move into a city

Counter-urbanisation

Figure 2 Causes and consequences of counter-urbanisation

Causes	Consequences
• People moving out of the city tend to be the most affluent and the most mobile. • They tend to be those with young children who think that the countryside will be a better place to bring them up. • The push factors from the city include traffic congestion, higher cost of living, perception of high crime, poor air quality, and the dream of the rural idyll. • Better road and rail links to city centres has enabled people to live further away from their places of work and commute easily. • Businesses with offices are now moving to more rural locations on the edge of cities where the land prices are cheaper and the quality of life for their workers can be better. • With improvements in high-speed broadband in rural areas, new industries can locate anywhere. • Similarly, improvements in telecommunications have enabled more people to work from home and still be in touch with their offices in cities.	• **Counter-urbanisation** creates dormitory villages, where residents work in the city during the day and only return to the rural area in the evenings. • House prices increase rapidly in relatively rural areas just outside urban centres, such as Hemel Hempstead outside of London. • Long-term residents of villages fear that the character of these settlements is changing. • Local people can be priced out of the market as wealthier city people buy up and renovate older properties, raising the profile of the settlement and subsequently the cost of properties there. This process is called gentrification.

Re-urbanisation

Figure 3 Causes and consequences of re-urbanisation

Causes	Consequences
• People are returning to live in the city, particularly in inner-city areas. • Government initiatives encourage people and businesses back into the city e.g. staff may be paid a premium for new jobs in deprived areas. • Grants have been available to retailers to take on derelict buildings. • Young people moving to the city for university and to find work need housing close to the amenities and institutions. • Gentrification has also helped to revive inner-city areas, where the housing offers easy access to work and entertainment in the city centre. • With better health care in the cities, older people who moved to rural areas when they retired may seek to return to the city for better access to hospitals.	• The redevelopment of inner-city urban areas creates new jobs and homes, which attracts people from outside to move there. • There can be a lack of affordable housing, which can lead to expensive new apartments being left empty. • Traffic congestion due to increased numbers of residents. • The process of gentrification might mean that working-class people are unable to buy or rent property in the city. • As the image of the city improves, more people are attracted to it.

Key term

Re-urbanisation: the use of initiatives to counter problems of inner-city decline

Revision activity

Using the information on this page and your own knowledge, copy and complete the table below showing the social, economic and environmental consequences of urban trends in the UK. Some consequences may fit into more than one column.

Urban trend	Consequence		
	Social	**Economic**	**Environmental**
Suburbanisation			
Counter-urbanisation			
Re-urbanisation			

Now test yourself

1 What encouraged the movement of people into the suburbs of cities?
2 What are the push and pull factors involved in counter-urbanisation?
3 Why are cities experiencing re-urbanisation?

Answers online

Challenges and ways of life in UK cities

Case study: A major city in the UK

Case study: Leeds, UK

Location and importance

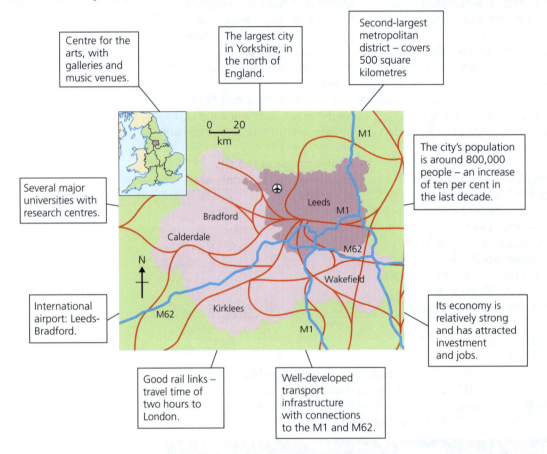

Centre for the arts, with galleries and music venues.

The largest city in Yorkshire, in the north of England.

Second-largest metropolitan district – covers 500 square kilometres

Several major universities with research centres.

The city's population is around 800,000 people – an increase of ten per cent in the last decade.

International airport: Leeds-Bradford.

Its economy is relatively strong and has attracted investment and jobs.

Good rail links – travel time of two hours to London.

Well-developed transport infrastructure with connections to the M1 and M62.

Bradford

Calderdale

Leeds

Wakefield

Kirklees

Figure 1 Location map of Leeds

Patterns of migration

- Around 17 per cent of the population are from black and ethnic minority communities.
- Residents who are non-UK born tend to settle in the Gipton and Harehills wards of the city.
- There was an influx of migrants, including 'new Commonwealth' immigrants from the Caribbean, during the 1950s. The city has a West Indian carnival every year.
- Pakistani and Indian communities were also well represented during the 1950s.
- A large Irish community established themselves in the early nineteenth century. They were spread through the city after the slum clearances, having initially settled in an area called 'the Bank'.
- After a second wave of immigration in the mid-twentieth century, the Irish community numbered over 30,000. They found work in labouring and manufacturing jobs.
- After the Second World War, the city welcomed Polish, Ukrainian and Hungarian refugees and, after the extension of the EU in 2004, new arrivals from Lithuania.
- Around 89 per cent of the population of Leeds were born in the UK.

- In 2013, the Office for National Statistics (ONS) estimated that there were between 6000 and 9000 new long-term immigrants in the city (net migration was around 1700).

Character of the city

- The city shows great diversity. Between 1991 and 2011, the ethnic minority population in Leeds doubled in size. The largest groups were Pakistani and Indian. Some ethnic minorities struggle to gain well-paid work. Pakistani communities show more preference for living in the same area of the city.
- The university influences the character of the city, with 30,000 students and 7000 staff at the University of Leeds alone. The number of young people creates demand for housing stock and they are important for the local economy.
- Leeds has a high proportion of young people; 18 per cent of the city are aged fifteen and under. Many retail and entertainment businesses cater for the teenage and young adult market. There is a thriving café culture.

Way of life

- **Industry**: the headquarters of ASDA and Danish company ARLA are in Leeds. It has a well-developed digital infrastructure that attracts new business. Many creative industries work from hubs based in old Victorian industrial buildings. M&S opened their first arcade in Leeds in 1884 and grew from there.
- **Sport**: the 2014 Tour de France started in Leeds, with millions of spectators lining the streets of the city. This was part of the Yorkshire tourism agency's bid to raise the profile of the city. There are football and rugby teams, as well as a cricket ground.
- **New developments**: new buildings are being erected along the waterfront and canal. The council plans to build affordable housing to tackle the issue of homelessness. Some businesses change their names during the week to provide different experiences for young people. There has been large-scale redevelopment, though not all developments have full occupancy.
- **Surrounding area**: the city sits close to the Yorkshire Dales and North York Moors National Parks and there are other Areas of Outstanding Natural Beauty (AONB) nearby.

Challenges

- **Studentification**: this term is used to describe a student community replacing the local community. This is true of South Headingley and Hyde Park. Studentification creates a demographic imbalance. There are often more pubs and takeaway restaurants in these areas, and crime tends to be high, particularly from antisocial behaviour. Pride in the community and the appearance of housing also tends to be reduced.
- **Housing availability**: due to studentification, areas have seen a rise in property prices. There have been new housing developments built along the Leeds-Liverpool Canal.
- **Transport issues**: there is demand for a better mass transportation system; fares are rising and there is a need for a well-developed tram system to reach more areas of the city.
- **Waste management**: every household produces 590 g of household waste per year, which is 166,100 tonnes of waste annually. Leeds council introduced plans to reduce waste and improve recycling by changing to fortnightly waste and recycling collection, and building a new recycling facility.
- **Social inequality**: the gap between the wealthiest and poorest residents is significant. In Holbeck, over 15 per cent of residents were on Jobseeker's Allowance and other benefits in 2015, while in Weetwood that figure was just 0.2 per cent. Leeds has been identified as having the third-highest levels of inequality of any city in the UK.
- **Loss of local businesses**: large retail developments such as Eastgate and Trinity have taken up local investment at the expense of local businesses, leading to the loss of independent, local businesses.

> **Key term**
>
> **Social inequality**: the extent to which people have unequal opportunities and rewards as a result of the position they occupy within the society; different groups, characterised by age, gender, 'class' and ethnicity, may have different levels of access to employment, education and health care

Sustainable initiatives

Plans for the South Bank

- Infrastructure and investment, including a new HS2 railway station
- Supports retail, leisure and financial services
- Cultural centre at the old Tetley brewery will support contemporary art
- Educational improvements linked to Leeds City College, particularly for vocational courses
- A new 3.5-hectare park and open space along the waterfront
- Over 300,000 square metres of development land available
- Creation of Holbeck Urban Village to improve the physical and social environment
- New pedestrian and cycle bridges
- Clarence Dock will become Leeds Dock and contain entertainment, restaurants and retail developments
- Water taxis and shuttle buses will connect the area, reducing carbon dioxide emissions

Exam practice

With reference to your chosen city case study, examine how ways of life can vary within one AC city.

[8]

ONLINE ☐

3 UK environmental challenges

Extreme weather in the UK

What affects the weather in the UK?

REVISED

It is important to know the difference between weather and climate.
- **Weather** – the day-to-day conditions of the atmosphere, for example temperature, rainfall and wind direction.
- **Climate** – the average weather conditions calculated over a 30-year period.

Several factors affect the weather in the UK.

Prevailing winds

The **prevailing** (or dominant) direction of **winds** affecting the UK is from the south-west. Winds blow from this direction most of the time. Travelling over the relatively warm Atlantic Ocean, this explains the moderate temperatures and high rainfall experienced in the UK.

Air masses

An **air mass** is a large body of air that transfers conditions of heat and moisture as it travels from its source area. Figure 1 shows that the UK is affected by a number of very different air masses. This explains our very changeable weather!

> **Key terms**
>
> **Prevailing wind**: the most frequent, or common, wind direction
> **Air mass**: a large parcel of air in the atmosphere; all parts of the air mass have similar temperature and moisture content at ground level

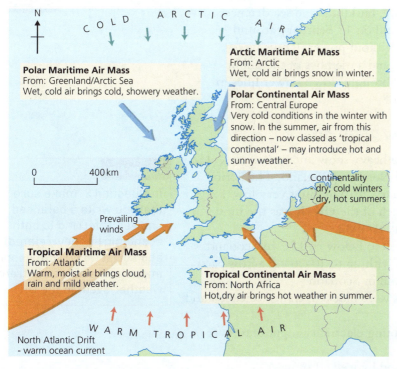

Figure 1 Air masses affecting the UK

> **Now test yourself**
>
> 1 What is the difference between weather and climate?
> 2 Which air masses are responsible for the following weather conditions in the UK:
>   (a) hot and sunny weather in the summer
>   (b) cold and wet weather, with some snow
>   (c) warm and moist weather, with cloud and rain?
> 3 What weather conditions area associated with the Polar Maritime Air Mass?
>
> **Answers online**
>
> TESTED

North Atlantic Drift

This is a warm ocean current that originates in the Caribbean and transfers warm conditions to the western side of the British Isles and beyond to western Scandinavia. The **North Atlantic Drift** explains why the western side of the UK tends not to experience severe winter weather.

Continentality

Continental interiors warm up and cool down more rapidly than maritime (coastal) areas. This explains why central Europe experiences cold winters and hot summers. It also accounts for the relatively low rainfall there as these areas are far away from the rain-bearing clouds off the Atlantic Ocean. Occasionally these continental conditions drift over the south-east of the UK, introducing very cold conditions in winter or hot weather in the summer.

> **Key term**
>
> **North Atlantic Drift**: a powerful ocean current responsible for maintaining warm conditions throughout the UK

How do air masses cause extreme weather?

REVISED ☐

Tropical maritime: winter storms 2014

In February 2014, the UK was battered by a succession of winter storms from the Atlantic, driven onshore by strong westerly winds. This was the stormiest period of weather experienced by the UK for 20 years.

- Strong winds and huge waves made conditions extremely dangerous around exposed coastlines – particularly in the south and west – and caused widespread transport disruption.
- The South West mainline railway was severely damaged at Dawlish, Devon, during the storm of 4–5 February, severing a key transport link to the South West for many weeks.
- Huge waves overtopped coastal flood defences and many coastal communities in Cornwall, Devon and Dorset experienced coastal flooding and damage to infrastructure, buildings and sea defences.
- Many trees were felled by the wind and on 12 February around 100,000 homes and businesses were without power. Roofs were damaged in Porthmadog, Gwynedd, and a member of the public was killed on 13 February after trees brought down power lines in Wiltshire.

> **Exam practice**
>
> 1 To what extent do air masses explain the changeable weather experienced in the UK? [6]
> 2 Describe the economic and social impacts of one named extreme weather event in the UK. [4]
>
> ONLINE ☐

Arctic maritime: heavy snow 2009-10

The Arctic maritime air mass can bring heavy snow and extremely cold conditions particularly to Scotland and Northern England. The period from mid-December 2009 to mid-January 2010 brought very low temperatures and heavy snowfall to much of the UK. It was the most severe period of winter weather since 1981/1982.

- Night-time temperatures regularly fell to below −10 °C in Scotland.
- Widespread snow fell across the UK with up to 10–20 cm across parts of England and Wales and up to 30 cm in Scotland.
- Transport was badly affected – roads were blocked, trains cancelled and airports disrupted.
- Ice brought down power lines disrupting electricity supplies to over 25,000 homes.
- Several people died in accidents caused by ice and snow.
- Farm animals across the UK were also severely impacted, particularly sheep in mountainous areas.

> **Exam tip**
>
> In Question 2, make sure that you write a balanced answer, referring to both economic (money-related) and social (to do with people) impacts of a named extreme weather event in the UK.

Extreme flood events in the UK

Extreme flood events

With the changing climate, flooding is becoming a major environmental challenge for the UK. The magnitude and frequency of future flood events are expected to be more severe in the future compared to the past.

In recent years, there have been a number of major flooding events in the UK:

- July 2007 – River Severn flooding inundated several towns and villages, including Tewkesbury.
- November 2009 – Cockermouth in Cumbria was severely affected by floods following torrential rain and several months of wet weather.
- January 2014 – vast areas of the Somerset Levels were flooded following incessant rainfall for several weeks.
- November/December 2015 – a series of winter storms brought flooding to many northern and western regions, with the Lake District suffering significant impacts.

Case study of a UK flood event caused by extreme weather conditions

Case study: Somerset Levels, 2014

The Somerset Levels is a vast flat agricultural area in south-west England lying between the Quantock and Mendip Hills, south of Bristol. In January 2014, an area equivalent to 65 km² became inundated by floodwater. Villages were cut off and many people and farm animals had to be rescued. The floods lasted for several weeks, causing immense disruption to people's lives and considerable environmental damage.

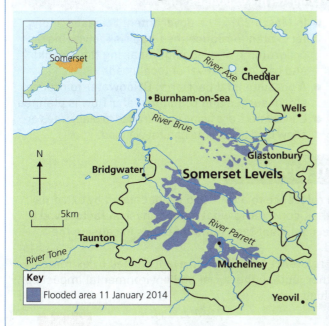

Figure 1 Location map of the Somerset Levels

Figure 2 Somerset floods, 2014

What were the causes?

Physical causes	Human causes
• This was the wettest winter period since 1910 – there were twelve major storms between December 2013 and February 2014, driven across the Atlantic Ocean by powerful high level winds (called the **jet stream**) • Strong tidal surges swept seawater into the river systems and prevented water from the land escaping to the sea – this led to flooding of the River Tone and River Parrett • Tidal surges deposited huge amounts of silt in the river channels, reducing their capacity to hold water • The land is extremely flat and low-lying, making it particularly vulnerable to flooding	• The River Parrett had not been dredged for some 20 years and its channel had become silted, reducing its capacity to cope with floodwater • Building on the floodplain (roads, houses) increased runoff into rivers • Drainage ditches used to drain the land for farming increased the speed of water flow into rivers **Key term** **Jet stream**: a narrow band of very strong wind currents that circle the globe several kilometres above the Earth

What were the effects?

Social	Economic	Environmental
• 600 homes were damaged • Some villages (e.g. Muchelney) were cut off for several weeks • Several roads were impassable, increasing journey times and leading to considerable inconvenience	• The cost to the Somerset economy was estimated to be between £82m and £147m • Farm animals had to be moved out of the area, fed expensive grain and sold • Businesses in the area lost money due to extra transport costs and lack of visitors	• Some 7000 hectares of productive agricultural land was underwater for several weeks • Some water became deoxygenated, affecting aquatic wildlife and leaving a black scum on the fields • Floodwater contained raw sewage, dead animals and toxic waste

What were the responses?

Short-term responses	Long-term responses
• The Environment Agency installed pumps sourced from as far away as the Netherlands to pump water back into the rivers • Emergency services, the Royal Marines and other volunteer organisations supported local people • As the floods receded, dredging of rivers began	• In March 2014, the government released a long-term plan to reduce the flood risk, involving a programme of dredging, repairing flood banks and raising the level of local roads • A possible tidal barrier at Bridgwater to reduce storm surges and regulate water flow is planned • Landowners are being encouraged to plant trees or allow land to become natural wetlands

Now test yourself

How did the following factors increase the risk of flooding in the Somerset Levels?
(a) prolonged period of heavy rain
(b) storm surges
(c) lack of river dredging

Answers online

TESTED ☐

Exam practice

1 With reference to a UK flood event caused by extreme weather, consider to what extent physical factors were more significant than human factors in causing the flood. [6]
2 With reference to a UK flood event caused by extreme weather, outline the social and environmental impacts of the flood. [4]

ONLINE ☐

Exam tip

It is important that you learn some specific details about your chosen UK flood event.
Make sure that you understand how it was caused by extreme weather.

Using and modifying environments and ecosystems in the UK

Land in the UK has been heavily modified to provide food, energy and water.

Providing food
REVISED

Farming

Farming in the UK has become very intensive, using machinery, technology and chemicals to maximise production. Whilst this has produced a high level of food security in the UK, environments and ecosystems have been significantly modified.

- Hedges have been removed to enable fields to become larger, accommodating big machines and maximising yields. This has had a significant effect on natural ecosystems by destroying habitats and eradicating shelterbelts and wildlife corridors.
- Intensive farming practices no longer involve fallow periods during which fields can recover. Soils become exhausted and more friable (dry and crumbly), making them prone to soil erosion by wind and water.
- Machinery can compress soils, leading to surface flooding.
- Chemicals used in fertilisers and pesticides can lead to water pollution.

Commercial fishing

Worldwide, over one billion people now rely on fish as their primary source of food. In the UK, commercial fishing utilises large sophisticated trawlers that are able to catch huge quantities of fish from the North Sea. The number of fishing boats has dropped by nearly a third since 1996.

- Commercial fishing has had significant impacts on fish stocks – landings of fish have more than halved since the 1970s, partly due to the decline of fish stocks.
- Some trawling practices have caught fish of all sizes, preventing smaller fish from growing to replace those that have been caught. This can result in very unbalanced ecosystems.
- Occasionally, unintended fish, such as dolphins, can be caught.
- Fishing boats often have to travel great distances to locate fish – this uses up diesel and increases the use of fossil fuels.

Providing energy
REVISED

Wind farms

Wind farms are an important source of renewable energy in the UK. In 2015, wind power contributed 11 per cent of the UK's electricity generation. Wind farms – large groups of individual turbines – can be situated on land or out to sea. Whilst they provide clean energy, they are controversial.

- Onshore wind farms tend to be located on high ground in the open countryside which some people consider to be unsightly, ruining views and the 'natural' landscape. Tourism can be affected when iconic landscapes – for example in the Lake District – are modified by the construction of wind farms.
- Some turbines are noisy to run, which may affect local people.
- Offshore wind farms are often less controversial, although they may interfere with bird migration and can disturb seabed ecosystems during construction.

Fracking

Fracking involves the injection of water and chemicals into rocks to extract oil and gas. Whilst it is a common practice in the USA (accounting for nearly 50 per cent of all gas and oil produced) it is highly controversial in the UK. In 2016, fracking in Lancashire was given the go-ahead by the government and drilling is expected to start in 2017.

> **Key term**
>
> **Fracking**: hydraulic fracturing; a controversial practice for extracting gas and oil from shale rocks

- There are concerns about the pollution of aquifers and soils as the injected fluids can contain toxic chemicals, including bromide, diesel and even hydrochloric acid.
- There are some concerns about fracking leading to minor earthquakes, although this may not be a major issue for the UK.

Providing water

REVISED

Water supply is an issue in the UK because most of the supply is in the north and west, yet most of the demand is in the south and east. About 80 per cent of water abstracted is used for energy or domestic use. Only 1 per cent is used in agriculture.

Two solutions to the problem of water supply are the construction of reservoirs to store water and the transfer of water from areas of high supply to areas of high demand. The Elan Valley water transfer scheme supplies Birmingham with water from over 100 miles away in Wales, via the Craig Goch dam and reservoir.

Reservoirs and water transfer schemes can affect environments and ecosystems.

Reservoirs	Water transfer schemes
• Reservoirs inundate landscapes, flooding agricultural land, natural ecosystems and even settlements • Reservoirs trap sediment, preventing its natural transportation downstream – this will affect ecosystems • Dam construction may damage the environment and harm ecosystems	• Water characteristics (chemical and physical) will be transferred between places, which may harm ecosystems as the chemistry of the water changes • River channels may become silted, leading to flooding and harming local ecosystems • Construction of pipelines and their maintenance may affect ecosystems

> **Revision activity**
>
> Construct a spider diagram to summarise the environmental and ecosystem modifications resulting from the provision of food, energy and water in the UK.

Exam practice

1 Explain how food production in the UK can modify ecosystems. [4]
2 Evaluate the extent to which developments in energy provision **or** water supply can harm the environment. [6]

ONLINE

> **Exam tip**
>
> In answering questions 1 and 2, be sure to focus on 'ecosystem' or 'environment' as you are required to do so. They are not the same, so make sure you are precise in your response.

Energy sources in the UK

Energy is used to power machinery and provide light and heat. The UK is fortunate in having a variety of sources of energy. There are two broad types of energy:

- **Renewable energy** – this is produced from energy sources that do not run out, such as solar (Sun), wind, water (hydro-electric) and tidal.
- **Non-renewable energy** – this is produced from energy sources that will eventually run out or become too expensive (economically or environmentally) to use. Sources of non-renewable energy include coal, natural gas and oil – so-called 'fossil fuels' because they were formed millions of years ago from the remains of living matter. Nuclear energy, which uses uranium, is also non-renewable.

> **Key term**
>
> **Renewable energy**: energy harvested from resources that are naturally replenished on a short timescale

Now test yourself   TESTED ☐

What is the difference between renewable and non-renewable energy? Give one example of each.

Answers online

Renewable energy sources   REVISED ☐

Renewable energy sources can be used over and over again without being used up. They are generally non-polluting. Apart from **biomass**, they do not directly involve the emission of harmful greenhouse gases.

The table below summarises the main types of renewable energy in the UK.

> **Key term**
>
> **Biomass**: the total mass of plants and animals in an ecosystem

Figure 1 Renewable energy sources

Renewable energy source	How does it work?	Importance in the UK
Biomass	Energy produced from organic matter. It includes burning dung or plant matter and the production of biofuels, by processing specially grown plants, such as sugar cane and maize.	Some biofuels are produced and used in transportation (about 3% of total road transport fuel). Biofuels and waste account for over 5% of the UK's electricity generation.
Wind	Turbines on land or at sea are turned by the wind to generate electricity. The UK is one of Europe's windiest countries!	In 2014, wind power accounted for just below 10% of the UK's electricity demand. Despite being unpopular, wind energy does have considerable potential for the future.
Hydro (HEP)	Large scale dams and smaller micro-dams create a head of water that can spin turbines to generate electricity	Large dams are expensive and controversial. Micro-dams are becoming popular options at the local level. HEP currently supplies just 1.4% of the UK's electricity production.
Geothermal	Water heated underground when in contact with hot rocks creates steam that drives turbines to generate electricity.	There are some small geothermal projects in the UK, e.g. in Southampton's city centre.

→

Renewable energy source	How does it work?	Importance in the UK
Tidal	Turbines within barrages (dams) constructed across river estuaries can use rising and falling **tides** to generate electricity	There are no existing tidal power barrages in the UK due to the high costs and environmental concerns. Tidal power could generate up to 10% of the UK's electricity. In the future, sites such as Swansea Bay and Bridgwater Bay might be developed for tidal power.
Wave	One method involves waves forcing air into a chamber, where it turns a turbine linked to a generator.	Portugal has installed the world's first wave farm, which started generating electricity in 2008. There are some experimental wave sites in the UK but costs are high and there are environmental concerns.
Solar	Most commonly, this involves photovoltaic cells mounted on solar panels which convert light from the sun into electricity	During the summer, solar power can generate considerable amounts of electricity. There are increasing numbers of solar farms in the UK and many homes have solar panels on their roofs. Solar power almost doubled during 2014.

Key term

Tides: changes in sea level as a result of the Moon; regular movements that occur every day

Now test yourself

TESTED

How is energy obtained from the following renewable sources?
(a) Hydro-electricity (HEP)
(b) Geothermal
(c) Biomass

Answers online

Non-renewable energy sources

REVISED

Non-renewable energy sources – in particular hydro-carbons such as oil and coal – emit large quantities of greenhouse gases and pollute the air. In the UK, there are three main types of non-renewable energy – coal, natural gas and nuclear.

- **Coal**: Coal used to be the main energy source in the UK. One hundred years ago, the industry employed over 3 million coal miners, mostly working deep underground. Today just a few thousand work in the industry and, in December 2015, the last deep-shaft pit at Kellingley, near Castleford, closed. Coal is now only extracted from huge opencast pits.
- **Natural gas**: Nearly half of the UK's electricity comes from natural gas. In the past, this has mostly come from the North Sea but today it is increasingly imported. By 2019, the UK is expected to import 69 per cent of the gas it requires for electricity generation.
- **Nuclear**: Nuclear power is a non-renewable energy source because it uses uranium as a raw material. It is an extremely important source of power in the UK. There are currently sixteen reactors generating about 20 per cent of the UK's total electricity. Figure 2 shows the location of the UK's nuclear power plants. Notice that they are located on the coast. This is because they require huge quantities of water for cooling. Some are close to ports, where imported uranium arrives. Several new reactors are planned in the future, including two each at Hinkley Point in Somerset and Sizewell in Suffolk. These should be up and running in the 2020s and will replace reactors reaching the end of their useful life.

Figure 2 Nuclear power plants in the UK

What is the energy mix in the UK?

REVISED ☐

The UK has a good **energy mix**, making use of several different sources of energy. Look at Figure 3. Notice that almost 50 per cent of the UK's energy comes from natural gas. Coal is still important despite the closure in 2015 of the last deep mine. Together, natural gas, coal and nuclear energy provide the majority of the UK's energy. Only a relatively small amount comes from renewable sources.

Key term

Energy mix: a measure of the different sources of energy in a given region

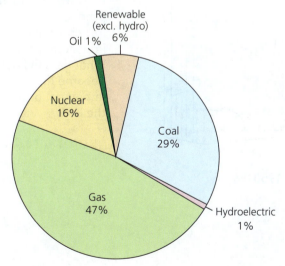

Figure 3 The UK's energy mix

Exam practice

1 What is meant by renewable energy? [2]
2 Study Figure 2. Explain the location of nuclear power plants in the UK. [4]
3 To what extent can the UK's energy mix (Figure 3) be considered sustainable? [6]

ONLINE ☐

Exam tip

In answering Question 2, make sure that you refer to the information in Figure 2 and that you **explain** (give reasons for) the location of the nuclear power plants. In Question 3, make sure that you focus on the command 'to what extent' – you need to express and justify your opinion.

Energy management in the UK
Changing patterns of energy supply since 1950

REVISED

Today, natural gas supplies almost 50 per cent of the UK's energy. Coal supplies nearly 30 per cent and nuclear 16 per cent. The UK has a reasonably balanced energy mix. This has not always been the case.

Figure 1 shows the production of primary fuels in the UK (1950-96). Primary fuels were either used directly to provide energy or indirectly to generate electricity.

In 1950:
- Almost all energy in the UK came from coal that was mined in the UK.
- Coal was used directly in industry and in the home to provide heating and hot water. It was used to fuel steam trains and to produce 'town gas', used in many homes for cooking.
- Coal generated almost all of the UK's electricity, apart from a small amount of hydro-electricity in Scotland.

Since the 1960s:
- Coal has declined erratically as the industry was hit by industrial action, mines were closed and employment fell.
- Oil increased rapidly due to production from the North Sea.
- Natural gas and nuclear increased, with several nuclear plants being constructed. Natural gas use has soared since the 1990s.
- Renewables increased only very slowly due to the high construction costs and the abundance of non-renewables.

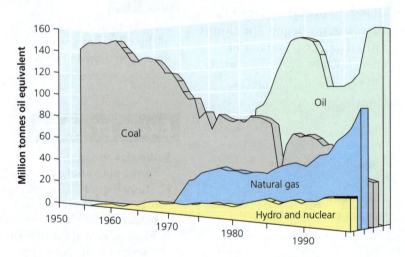

Figure 1 Production of primary fuels in the UK, 1950-96

Now test yourself

Study Figure 1.
1. What was the production of coal in 1960?
2. What date was there a sudden fall in coal production?
3. Describe the changes in hydro and nuclear production.
4. In 1996, what was the most important primary fuel?
5. What date did natural gas become the second most important primary fuel in the UK?

Answers online

TESTED

The role of the government and international organisations in the UK's changing energy patterns

Many of the changes that have taken place since the 1950s reflect government decisions and the role of international organisations.

- In 1961, coal accounted for 80 per cent of the UK's energy supply. Following a prolonged miners' strike in 1984–85 over mine closures and job losses, coal production declined rapidly. It had become increasingly expensive to mine and alternative energy sources were becoming available, particularly oil and gas from the Middle East.
- In 1974, the international organisation OPEC (Organisation of Petroleum Exporting Countries) quadrupled the price of oil on the world markets. Fortunately the UK had started to develop its own North Sea oil and gas, with the first oil being produced in 1975. Multi-national oil companies were involved in exploration and production with the UK government taking money through taxation.
- In the early 1990s, the EU's 'Gas Burn' directive was repealed. Up until then, the directive had imposed restrictions on the use of gas. Once repealed, the gas industry was able to grow rapidly.

- The EU and the United Nations have responded to the challenges of global climate change by setting limits on carbon emissions. This has led to reductions in the use of hydro-carbons (particularly coal and oil) in favour of developing renewable sources of energy, such as wind and solar. Natural gas is the favoured hydro-carbon, as it has relatively low carbon and sulphur emissions. Many of these developments have been supported by the UK government, which has a target of reducing carbon emissions in the UK by 60 per cent by 2050.
- In the 2007 White Paper 'Meeting the Energy Challenge', the government recognised the need to build new electricity power stations to replace those built in the 1960s and 1970s. This accounts for plans to build new nuclear reactors, for example at Hinkley Point in Somerset using foreign investment from China.

> **Revision activity**
>
> Construct a spider diagram to summarise the roles of the government and international organisations in the UK's changing energy patterns.

Changing patterns of energy demand since 1950

REVISED ☐

Energy consumption in the UK has risen since the 1950s due to:

- growth in the UK's population
- increasing energy demands in the home (for example central heating, electrical appliances)
- increased demand from industry and transport (massive increase in car ownership).

In the period 1970–2000, energy consumption increased by about 15 per cent. Since 1990, it has increased by about 1 per cent per year and after 2000 it has actually started to decline.

Declining energy demand in the UK

Between 2005 and 2011, overall energy consumption in UK homes fell by more than 25 per cent. This coincided with a sharp increase in the cost of energy, together with a number of other factors:

- Government insulation policy. Under CERT (Carbon Emissions Reduction Target), one of the large-scale energy efficiency schemes, 5.3 million homes received free loft insulation and 2.6 million got their cavity walls filled. The scheme finished in 2012.
- Since 2007, buildings in England and Wales have to undergo Energy Performance Certification before they are sold or let. This encourages homeowners to be more energy efficient.
- New standards set by the government. Since 2005, all new boilers have at least a B rating. National Grid thinks that this could account for more than 40 per cent of the overall drop in household demand.
- Environmentalism. Awareness of our carbon footprint and efforts by the Energy Saving Trust and other international organisations, such as the United Nations, have encouraged people to use less energy.

The role of the EU

The EU has a target of reducing carbon emissions by 40 per cent by 2030. It too is encouraging reductions in energy use:

- EU 'energy label' details the energy efficiency of products, such as light bulbs and washing machines. Products are rated from A–G, with A being the most efficient and having a green colour. From 2017, there will be 7 categories from A★★★-D.
- In 2011, the EU adopted the 'Energy Efficiency Plan', which encourages the construction of more energy efficient houses, encourages industry to produce more energy efficient products and promotes the use of 'smart meters'.

Now test yourself

Explain how the following factors have led to a reduction in energy demand:
(a) loft insulation
(b) Energy Performance Certification
(c) EU energy labels.

Answers online

TESTED ☐

REVISED ☐

Sustainable energy solutions

Sustainable energy solutions ensure the long-term availability of energy for future generations. This achieves national and community security and avoids damage to the environment.

National strategies

The UK government wants to create a low-carbon, sustainable future that helps to address climate change. There are four key aspects of this vision for a sustainable future.

1 **Increase the contribution of renewable sources.** The government's Renewable Energy Strategy (2009) identified a target of 15 per cent of the UK's energy to come from renewable sources by 2020. In 2013, it was just over 5 per cent. A target of 500,000 jobs will be created in the renewable sector by 2020 and communities will be encouraged to invest in micro-schemes to generate electricity.

2 **Encourage energy saving and conservation.** Grants have been available for loft insulation and all homes need to have an energy efficiency survey before being sold or rented. The EU requires energy labelling for appliances such as fridges. Technology is being encouraged to develop ever more efficient household appliances.

3 **Develop nuclear energy.** Whilst not strictly renewable, nuclear power uses very small amounts of raw material and some of this can be re-processed for further use. It also has very low carbon emissions.

Key term

Sustainable (sustainability, sustainable development): this approach places emphasis on improving the current quality of life but still maintaining resources for the future; it is a balance of providing social, economic and environmental benefit in the long term

Revision activity

Use a simple flow diagram to explain carbon capture and storage (Figure 2).

Renewable energy will not address all of the UK's energy needs and nuclear energy represents a long-term alternative that is reasonably sustainable.

4 **Develop carbon capture and storage.** Technology is now available to capture carbon from power stations and store it underground within rocks or aquifers (Figure 2).

Recent developments in sustainable energy solutions include:

- Between 2010 and 2013, the Department of Energy and Climate Change recorded £31 billion of private-sector investment in renewable electricity generation, supporting over 35,000 jobs.

Figure 2 Carbon capture and storage

Labels in figure: 1. Mining of fuel; Gas field; 2. Coal or gas-fired power station with CO_2 capture plant; 3. CO_2 transport by pipeline; 4. CO_2 injection; saline aquifers; unmineable coal seams; depleted oil and gas fields; 5. CO_2 storage sites

- In 2014, £10 billion was invested in renewable electricity generation, which included the Pen y Cymoedd wind project in Wales and the world's largest gasification plant on Teesside.

However, in 2015, the government published plans to cut subsidies for renewables such as solar and wind. This may lead to job losses in the industry and the closure of solar panel companies. Some new wind and solar projects may be in jeopardy.

Local strategies

In the UK there are many local sustainable energy projects involving private individuals and small communities. These include wind turbines, micro-hydro schemes and anaerobic digesters. Whilst most schemes involve private funding, financial and technical support is available from the government and other organisations.

For example, PlanLoCaL is a support organisation developed by the Centre for Sustainable Energy in 2009. It receives financial support from the government's Department for Communities and Local Government and Department of Energy and Climate Change. PlanLoCaL provides advice and support for communities wishing to develop sustainable energy solutions.

Case study: anaerobic digestion, Silloth, Cumbria

In 2011, an anaerobic digestion system was installed at Dryholme Farm near Silloth, an isolated farming community on the Cumbrian coast in north-west England. The digester uses farm slurry and silage made from locally grown grass and maize to generate enough electricity to power 4000 homes. The cost of the project was about £4m. Money came from a variety of sources including government grants, loans and private investments.

Anaerobic digestion works much like a cow's stomach! In the absence of oxygen (hence the term 'anaerobic') bacteria is used to break down the slurry and silage, creating a methane-rich biogas (Figure 3). The methane is burned as a fuel to generate electricity which is sold to the National Grid to make a profit.

In addition to the electricity generated by the plant, the waste organic matter (digestate) forms a valuable liquid fertiliser that farmers can spread onto their land. Heat is also produced by the plant which can be used locally.

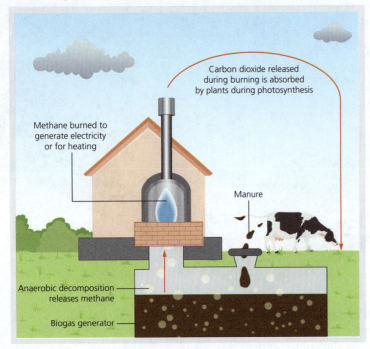

Methane burned to generate electricity or for heating

Carbon dioxide released during burning is absorbed by plants during photosynthesis

Manure

Anaerobic decomposition releases methane

Biogas generator

Figure 3 How anaerobic digestion works

The development of renewable energy in the UK

REVISED

In the 1950s, almost all of the UK's energy supply was coal. Only a very small amount of energy came from another source – hydro-electric power (HEP).

- In 1967, the world's first pumped storage HEP station at Cruachan Loch in Argyle and Bute, Scotland began operations. Water passes from Cruachan Loch through generating turbines to a lower lake, Loch Awe. At night, when demand is less, water is pumped back to the top lake so that it can be used again. Cruachan Loch is one of four pumped storage schemes in the UK.
- In the 1970s, a sudden rise in oil prices and miners' strikes sparked government research into renewable energy. At first, this involved investigating wave power. After several experiments, this was considered too expensive.
- In 1986, Southampton started to use geothermal heat to pump hot water through a district heating system.
- By the late 1980s, several wind turbines had been constructed and, in 1991, the first wind farm began operating at Delabole in Cornwall.
- In 1990, renewables contributed less than 2 per cent of the UK's electricity generation. By 2015, this had increased to 25 per cent. This rapid increase reflects government targets to reduce greenhouse gas emissions and the provision of generous subsidies.

Figure 4 shows the steady increase in renewable energy since 2006. Notice that wind accounts for almost half of the electricity generated by renewables.

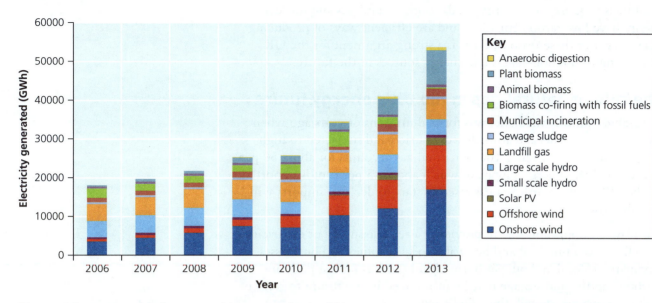

Figure 4 Recent trends in renewable energy in the UK

What are the impacts of wind farms on people and the environment?

Wind farms are highly controversial. Onshore wind farms are often planned to be sited in areas of open countryside on high ground so that they can catch the strongest wind. These same areas are much cherished for their natural beauty and local residents are often unhappy about any proposed construction. Whilst some people are concerned about the operating noise, most concerns relate to visual pollution and the possible impact on tourism.

Kirkby Moor Wind Farm, Cumbria

In 2015, local councillors objected to plans by RWE Innogy Ltd to replace twelve existing 42-metre turbines with six huge turbines, 115m tall. They were concerned about the impact of the new turbines on the quality of the Lake District landscape and feared that tourism would be affected.

Many people travel to the Lake District to see the same views that inspired artists and writers such as William Wordsworth. Local people are concerned that visitors will stop visiting if these views are spoiled. This would have a harmful impact on the economy of the area and would affect people's livelihoods.

Navitus Bay offshore wind farm

In 2015, ministers rejected an ambitious £3.5 billion plan to construct over 190 wind turbines, each nearly 200 metres high, in the English Channel near the Isle of Wight. They concluded that the offshore wind farm would harm the views from Dorset's Jurassic Coast, putting at risk its status as England's only Natural World Heritage Site. Tourism could suffer and this would have a negative impact on the economy of the region.

Revision activity

Draw a timeline to show the developments of renewable energy in the UK.

Now test yourself

Suggest **two** arguments in favour and against the development of wind farms.

Answers online

TESTED

The UK's future energy supply

Despite the recent growth in renewables, the UK is still dependent on fossil fuels to satisfy our energy needs. Both natural gas and nuclear energy have low carbon emissions and are efficient ways of producing electricity. For these reasons, there is a strong argument for the UK continuing to use non-renewable energy sources in the future.

The role of natural gas in the UK's energy future

Natural gas (mostly methane) was formed millions of years ago when the remains of plants and animals were buried and subjected to high temperatures and pressure. Today, natural gas accounts for about 50 per cent of the UK's energy supply, of which about 45 per cent is imported. As the UK's own North Sea gas reserves start to decline, imports will probably rise.

Gas plays an important role in providing a reliable supply of electricity as well as providing heat and hot water to UK homes and businesses. Compared to coal and oil, gas burns cleanly so it is far less polluting. Carbon capture and storage may be introduced in the future to reduce carbon emissions, though this will be expensive.

In terms of energy security, natural gas is reasonably secure. The UK has its own North Sea reserves and imports come from a range of different countries including Norway, Qatar and the Netherlands.

The fracking debate

Large quantities of oil and gas are trapped deep underground in shale formations. One method of extracting both oil and gas from shale involves hydraulic fracturing, commonly known as '**fracking**'. This involves pumping water, sand and chemicals into the shale, under high pressure. It fractures the rock and enables the oil and gas to escape and be extracted.

Fracking is used in the USA and elsewhere in the world but it is controversial. There are concerns about minor earthquakes and the pollution of groundwater. In 2016, fracking was given the go-ahead in the UK in Lancashire.

The nuclear debate

- Nuclear energy currently accounts for about 16 per cent of the UK's energy supply and generates about 19 per cent of the UK's electricity.
- Many of the country's nuclear power stations were constructed in the 1960s and 1970s and some are now reaching the end of their lives.
- The UK's most recent nuclear power station – Sizewell B in Suffolk – began operating in 1995, over 20 years ago!

Nuclear power stations are extremely expensive to build, yet they do offer relatively efficient and sustainable power for the future. In 2015, the Chinese announced that they would invest in the UK's nuclear programme by supporting new developments at Hinkley Point in Somerset, Bradwell in Essex and Sizewell.

Case study: Hinkley Point C power station, Somerset

The new nuclear power station, Hinkley Point C, will be the third to be sited on the north Somerset coast.
- Hinkley Point A was completed in 1965 but was de-commissioned (stopped producing electricity) in 2000.
- Hinkley Point B was commissioned in 1976 and is expected to remain operational until 2023. It is operated by the French energy company EDF.

In 2015, China agreed to fund about a third of the cost of constructing a new nuclear power plant at Hinkley Point. The total cost of the project is estimated to be £18 billion. The plant will be run by the French company EDF and is expected to start generating electricity in 2025.

The high costs involved mean that electricity will cost about twice the current amount and this could lead to price rises for consumers. However, an estimated 25,000 jobs will be created and the government expects the power station to generate enough electricity to supply over 5 million homes.

Now test yourself
TESTED

1 What is the name of the new nuclear power station in Somerset?
2 Which country is contributing a third of the cost?
3 How many jobs will be created?
4 The new power station will supply electricity to 1 million homes. True or false?

Answers online

What are the factors affecting the UK's future energy supply?
REVISED

Figure 5 outlines the economic, political and environmental factors affecting the future of the UK's energy supply.

Figure 5 The economic, political and environmental factors affecting the UK's energy supply

Economic	• The high cost of building new nuclear and gas-fired power stations, as well as the decommissioning of old power stations. • New power stations may result in more expensive electricity, which could hit businesses and consumers. • With North Sea supplies starting to dwindle, it will become increasingly expensive to extract oil and gas. • High cost of constructing renewable energy alternatives, such as wind farms, tidal barrages and HEP. • For individuals and local communities, small-scale renewable projects have to be cost-effective. Most require grants or loans.
Political	• Fracking could become a political issue with political parties each adopting a different stance. • Imports of natural gas, ensuring energy security by having a wide import base involving agreements with stable nations. • To what extent should foreign countries own, operate or invest in UK energy, such as the Chinese investing in new nuclear power stations in Somerset? • Will the government continue to support and encourage the renewable sector with grants and subsidies? The solar industry may suffer as its subsidy has been cut.
Environmental	• The UK is committed to reducing carbon emissions, so will need to ensure low carbon fuels are used to provide energy in the future. • Fracking may have environmental impacts, particularly involving pollution of groundwater aquifers. • Many people are concerned about nuclear power – the radioactive waste and the dangers of a radioactive leak. Terrorism, too, is a concern. • Environmental concerns may prevent expansion of the renewable sector by preventing the construction of wind farms and solar farms.

Exam practice

1 Study Figure 1. Describe the trends in the production of primary fuels in the UK between 1950 and 1996. [4]
2 Examine the role of the government and international organisations in the UK's changing energy patterns. [6]
3 Outline the strategies for sustainable use and management of energy at the local scale. [4]
4 Evaluate the extent to which non-renewable energy could and should contribute to the UK's future energy supply. [6]
5 Discuss the environmental factors affecting UK energy supply in the future. [6]

ONLINE

Exam tip

In answering question 1, make sure that you describe overall trends as well as **anomalies** (spikes and troughs). Use figures and dates from the graph. In question 4, be careful to focus on **non-renewable** energy only and make sure that you do 'evaluate' by weighing up the pros and cons.

Revision activity

Use a spider diagram to summarise the economic, political and environmental factors affecting the UK's future energy supply.

Key term

Anomalies: data values that don't match the pattern of a sample

4 Ecosystems of the planet

The components of ecosystems

What are ecosystems?

Ecosystems are natural areas in which plants, animals and other organisms interact with each other and the non-living elements of the environment. Plants are technically known as **flora** and animals are called **fauna**. The living elements of an ecosystem are known as **biotic**; the physical, non-living parts are known as **abiotic**.

Figure 1 Biotic and abiotic elements of an ecosystem

Biotic	Abiotic
● Animals, including insects, birds and mammals ● Plants, including trees, grasses, mosses and algae, which provide food and shelter for animals ● Micro-organisms, such as fungi, which break down dead plants and animals, releasing nutrients into the ecosystem ● Humans are also living parts of an ecosystem	● Sunshine and rain, which are needed for photosynthesis – this is dependent on the weather and climate ● Soils, which store water and carbon nutrients that plants can use ● Rocks, which help in the formation of soils; weathering releases nutrients stored in rocks into the ecosystem

Key terms

Ecosystem: an area where living organisms and non-living elements interact with each other
Flora: another term for the plants in an ecosystem
Fauna: another term for the animals in an ecosystem
Biotic: the living elements of an ecosystem
Abiotic: the non-living elements of an ecosystem
Interdependence: the reliance of every form of life on other living things and on the natural resources in its environment, such as air, soil and water

How are ecosystems 'interdependent'?

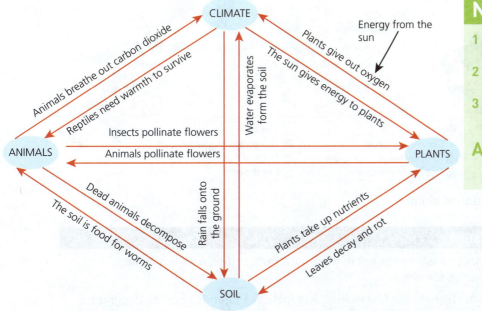

Figure 2 Interdependence in an ecosystem

Now test yourself

1 Explain how plants are dependent on the soil.
2 Explain how animals are dependent on the climate.
3 Explain how the climate is dependent on plants.

Answers online

The distribution and characteristics of ecosystems

What is the global distribution of the major biomes?

REVISED

Key term

Biome: large-scale ecosystems that are spread across continents and have plants and animals that are unique to them

The world contains eight major **biomes**. They are mostly terrestrial (on land) but some, like coral reefs, are marine (in the sea). Each biome has unique climatic conditions that create distinctive environments for plants and animals to adapt to and survive in.

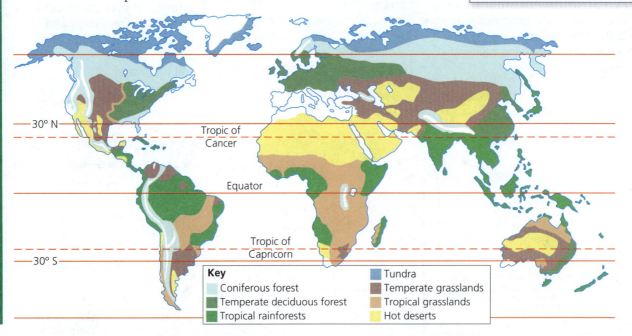

30° N

Tropic of Cancer

Equator

Tropic of Capricorn

30° S

Key
Coniferous forest
Temperate deciduous forest
Tropical rainforests
Tundra
Temperate grasslands
Tropical grasslands
Hot deserts

Figure 3 The global distribution of biomes

Biome	Location
Polar regions (Antarctica – South)	Antarctica is a continent in the South Pole region covered by an immense ice shelf.
Polar regions (the Arctic – North)	The Arctic is in the North Pole and includes islands such as Greenland and the northern parts of countries such as Russia and Canada.
Coral reefs	Coral reefs cover less than one per cent of the world's ocean. In total, 109 countries have coral reefs in their waters. Figure 1 on page 76 shows the location of coral reefs.

→

Biome	Location
Tropical rainforests	Around the Equator and within the tropics Amazon River Basin, South America South East Asia and Queensland, Australia
Tropical grasslands (savannah)	Between 5° and 30° north and south of the Equator, in central parts of continents Most of central Africa, surrounding the Congo Basin Northern Australia Brazilian Highlands
Hot deserts	Around 30° north and south of the Equator, typically on the west coast of continents around the tropics Sahara in northern Africa Mojave in North America
Temperate grasslands	Between 40° and 60° north and south of the Equator, in central parts of continents Plains of North America Veldts of Africa Steppes of Eurasia
Temperate forests	Between 40° and 60° north and south of the Equator Eastern North America, Western Europe (including the UK) and New Zealand

What are the characteristics of the major biomes?

REVISED

Figure 4 Key characteristics of the major biomes

Biome	Climate	Flora and fauna
Polar regions (Antarctica – South)	● Long, cold winters and short cool summers. Covered in snow and ice throughout the year. Spends half the year in darkness ● South Pole winter temperatures vary from −62°C to −55°C ● The weather in Antarctica is kept within the continent due to **circumpolar winds** travelling around the continent ● Antarctica has an average height of 2300 metres (note that temperature decreases with altitude)	● Only 1% of the continent is ice free so plant life is less plentiful than in the Arctic ● Around 100 species of moss and 300–400 species of lichen grow on exposed rocks ● Large numbers of penguin species such as gentoo, emperor and Adélie ● Fur seals and elephant seals ● Killer whales and minke whales ● Both poles have rich seas due to large volumes of phytoplankton
Polar regions (the Arctic – North)	● Long, cold winters and short cool summers. Covered in snow and ice throughout the year. Spends half the year in darkness ● North Pole winter temperatures vary from −46°C to −26°C ● The sea in the Arctic does not fall below −2°C, causing the Arctic region to stay warmer than Antarctica ● Relatively warm weather from the south travels north into the Arctic region by the Gulf Stream	● Approximately 1700 species of plant ● Mosses, grasses and alpine-like flowering plants ● Low shrubs, reaching heights of around two metres ● Treeless due to the permafrost ● Land mammals including polar bears, wolves, foxes and reindeer ● Sea mammals including walruses and whales ● Some animals are able to migrate southwards during the winter months

→

Biome	Climate	Flora and fauna
Coral reefs	• For coral reefs to grow, they need warm water all year round with a mean temperature of 18 °C; the water needs to be clear and shallow (no deeper than 30 metres) to ensure that there is enough sunlight for photosynthesis	• Sea grasses, such as turtle grass and manatee grass, are commonly found in the Caribbean Sea. Sea grasses are flowering plants that provide shelter and habitat for reef animals, and food for **herbivores** such as reef fish • Reefs are made up of thousands of coral polyps. They look like plants but are, in fact, an animal related to the jellyfish. Each polyp is 2–3 cm in length and feeds on plankton. They make their own mineral skeleton from calcium carbonate. Other species found in coral reefs include parrot fish, starfish, clams, eels, dugongs, crustaceans and sponges
Tropical rainforests	• Hot and wet climate with no seasons • Monthly temperatures are high throughout the year, between 26 °C and 28 °C • Annual rainfall is often over 2000 mm; a thundery downpour happens most afternoons	• 15 million plant and animal species have been identified • Vegetation exists in distinct layers, from the highest emergents to the canopy, under canopy and shrub layer • Buttress roots support tall trees and vine-like plants called lianas grow between trees • Animals species include toucans, monkeys, chameleons and frogs; the poison dart frog is brightly coloured to warn off predators
Tropical grasslands (savannah)	• A longer dry season and a shorter wet season • Wet season arrives when the Sun moves overhead (ITCZ), bringing with it a band of heavy rain; 80% of the rain falls in the four or five months of the wet season • During the dry season rainfall is as low as 10 mm • Temperatures are high throughout the year but with a greater range than the rainforest; daily temperatures reach 25 °C	• Tall and spiky pampas grass grows quickly to over 3 metres • The baobab tree has large swollen stems and a trunk with a diameter of 10 metres to store water; it has a small number of leaves to reduce water loss via **transpiration** • 40 species of hoofed animals, e.g. antelope • Grazing species such as elephants, giraffe and wildebeest • **Carnivores**, such as lions and hyenas, stalk herds
Hot deserts	• Daytime temperatures of 36 °C but temperatures at night can fall to −12 °C due to the lack of insulating cloud cover • Annual precipitation is around 40 mm	• Most plants are **xerophytic** (adapted to survive a lack of water) • Cacti and yucca plants absorb water and have roots near the surface • Camels have humps to store water and fat, and long eyelashes to keep the sand out • Other species that have adapted to the extreme dryness are meerkats and sidewinder snakes
Temperate grasslands	• Summers are very hot, reaching over 38 °C, while winters are very cold, plummeting to as low as −40 °C • Average rainfall varies from 250 mm to 750 mm with 75% of rain falling in the summer season • Melting snow helps the start of the growing season	• Trees and shrubs struggle to grow but some, such as willow and oak, grow in river valleys • Tussock grasses grow to heights of 2 metres while buffalo and feather grasses grow more evenly to 50 cm. • Burrowing animals such as gophers and rabbits, and large herbivores such as kangaroos and bison • Carnivores including coyotes and wolves, as well as eagles

→

Biome	Climate	Flora and fauna
Temperate forests	<ul><li>Due to the tilt of the Earth and the angle of the Sun's rays, there are four distinct seasons</li><li>Summers are warm and winters are mild</li><li>Rainfall occurs throughout the year and ranges from 750 mm to 1500 mm with an annual average temperature of 10°C</li></ul>	<ul><li>**Deciduous** trees shed their leaves in the winter</li><li>Oak trees reach heights of 30–40 metres; other tree species include ash, birch and hawthorn</li><li>Grass and bracken grow on the forest floor</li><li>Some animals migrate or hibernate in winter, e.g. black bears in North America</li><li>Common species include squirrels, owls, pigeons, rabbits, deer and foxes</li></ul>

Key terms

Circumpolar winds: flows of air around the Earth's poles

Herbivore: an animal that feeds on plants

Transpiration: the process by which plants lose water vapour through their leaves

Carnivore: an animal that eats other animals

Xerophytic: a type of plant that can survive on very little water

Deciduous: trees that shed their leaves during winter to retain moisture, also known as broadleaved trees

Now test yourself

TESTED

1 Give **five** examples of how plants and animals from any biome have adapted to survive in their environment.
2 Describe how temperature and rainfall varies between each of the biomes.
3 How does the position of the ITCZ affect patterns of rainfall in tropical grasslands?
4 State two places in which you would find:
   (a) hot deserts
   (b) tropical rainforests.
5 Why is there such a large range of temperatures across 24 hours in the hot desert?

Answers online

Exam practice

Looking at Figure 5, describe the climate of Iqaluit. [2]

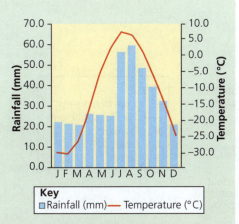

Figure 5 Climate graph of Iqaluit, Canada (Arctic)

ONLINE

Exam tip

For Question 1, remember that 'describe' means that you must say what you see on the graph. You do not need to give any reasons for the climate.

The world's major tropical rainforests

Where are tropical rainforests located?

REVISED

Figure 1 shows that tropical rainforests are found in a broad belt through the Tropics. This biome contains 50 per cent of all known plant and animal species. It is characterised by high rainfall (over 2000 mm a year) and high temperatures (averaging about 27 °C) throughout the year.

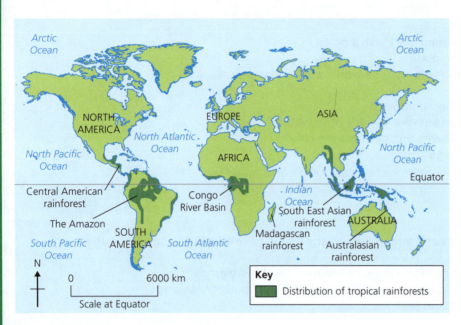

Figure 1 Location of tropical rainforests

What are the processes in a tropical rainforest?

REVISED

The warm and humid conditions promote natural processes such as plant growth and decay. Figure 2 shows the nutrient cycle in tropical rainforests.

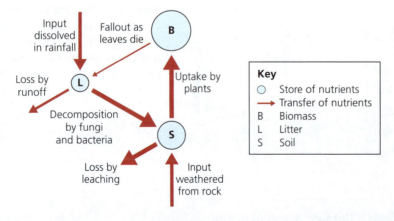

Figure 2 Nutrient cycle in the tropical rainforest ecosystem

- Nutrients (plant foods) are stored in the biomass, soil and **litter**. The size of the circle indicates the relative amount of nutrients stored.
- Most nutrients are stored in the biomass. Few nutrients are stored in the soil and litter due to the rapid uptake by plants and rapid decomposition involving fungi and bacteria, which thrive in warm and wet conditions.
- Nutrient inputs include precipitation and weathering. Leaching – the dissolving and carrying away of nutrients by rainwater – is a significant loss from the system.
- Nutrients are transferred between the stores by natural processes such as leaf fall and decomposition. The arrows show these transfers and the thickness of the arrows indicate the relative amounts of nutrients transferred.
- Most of the arrows are large, demonstrating the rapid transfer of nutrients between the stores. This is due to the ideal climatic conditions (wet and warm) for transfer processes. It explains why the nutrient stores in the soil and the litter are so small.

> **Key term**
>
> **Litter**: the total amount of organic matter, including humus (decomposed material) and leaf litter

Now test yourself

TESTED ☐

Study Figure 2.
1 What are the three nutrient stores in order of size (largest to smallest)?
2 Explain the relative size of the litter store.
3 What are the two inputs to the soil store?
4 What are the two outputs from the litter store?
5 Why is 'uptake by plants' a major transfer process?
6 What is leaching and why is it a significant process?

Answers online

Exam practice

Study Figure 1. Describe the distribution of tropical rainforests in the world. [4]

ONLINE ☐

> **Exam tip**
>
> When describing the distribution, refer to Figure 1, giving specific locations in your answer. Do not write about the **reasons** for the distribution.

The world's major coral reefs

Where are coral reefs located?

Figure 1 shows that coral reefs are mostly found in the Tropics, with significant concentrations in the Caribbean, the Indian Ocean and the Pacific Ocean, particularly in South East Asia and Oceania. Described as the 'rainforest of the seas' on account of their remarkable biodiversity, coral reefs require three main conditions for their development:

- Temperature – corals thrive in warm water, ideally 23–25 °C.
- Light – corals feed on algae which require light to photosynthesise. This explains why corals are found in relatively shallow water where light penetrates.
- Clear water – sediment in water impedes feeding and can kill corals, which explains why they are largely absent from river mouths.

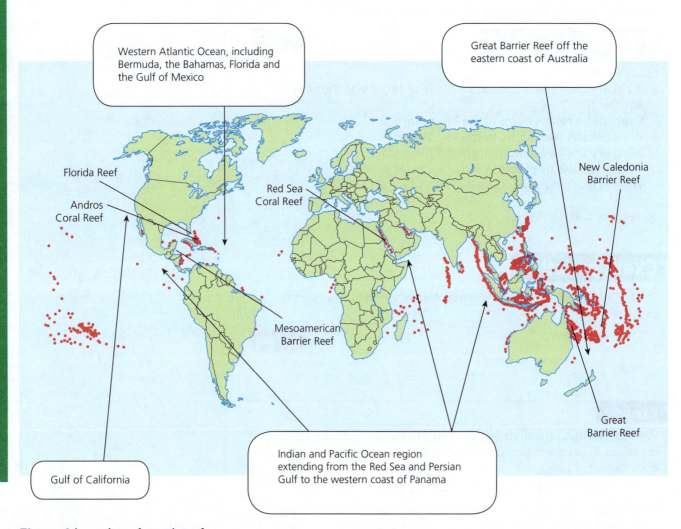

Figure 1 Location of coral reefs

What are the processes in a coral reef?

Figure 2 shows **nutrient cycling** in a coral reef ecosystem. Corals often exist in nutrient poor water so recycling is essential to ensure survival.

Key term

Nutrient cycling: a set of processes whereby organisms extract minerals necessary for growth from soil or water, before passing them on through the food chain, and ultimately back to the soil and water

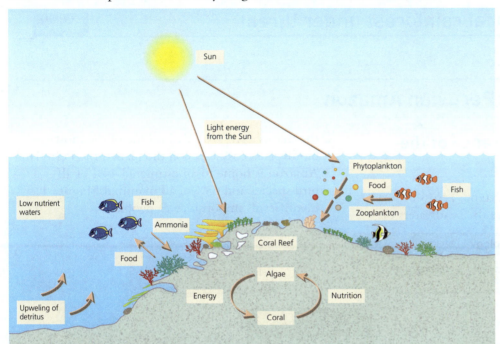

Sun

Light energy from the Sun

Low nutrient waters

Fish

Ammonia

Food

Phytoplankton

Food

Fish

Zooplankton

Coral Reef

Algae

Energy

Nutrition

Upweling of detritus

Coral

Figure 2 Simplified coral reef nutrient cycle

- Microscopic plant-like algae called zooxanthellae live within the tissues of the coral polyp, converting light from the sun into energy that can be used by the coral.
- The zooxanthellae benefit from waste nutrients provided by the corals. Nitrogen – the most important nutrient in the coral ecosystem – is passed back and forth between the coral and the zooxanthellae.
- Corals also obtain nutrients by feeding on zooplankton (which survive by consuming phytoplankton, the sea's primary producer that converts light from the sun into energy), bacteria and edible detritus from the ocean floor.
- Fish excrete ammonia (a dissolved form of nitrogen) into the water and this can be absorbed by corals and algae.

Now test yourself

Study Figure 1. Describe the distribution of coral reefs.

Answers online

TESTED

Exam practice

Describe the processes responsible for recycling nutrients in the coral reef ecosystem. [4]

ONLINE

Exam tip

Make sure you show clear connections between the **processes** and the **recycling of nutrients**.

Bio-diverse ecosystems under threat from human activity

Case study: tropical rainforest under threat

Case study: the Peruvian Amazon

What is the importance of the Peruvian Amazon?

The Peruvian rainforest extends across 60% of Peru and mostly comprises the headwaters of the massive Amazon basin. One of the most biodiverse ecosystems on the planet, the Peruvian Amazon is home to an estimated 44% of all bird species and 63% of all mammals! Figure 1 identifies its importance to people.

Archaeology – remains of ancient civilisations such as the Chachapoya (the so-called 'Cloud Forest People') exist within the rainforest

Biodiversity – almost 3000 known species of amphibians, birds, mammals and reptiles

The importance of the Peruvian Amazon

Indigenous tribes – many traditional tribes live in remote parts of the rainforest, living simple but sustainable lives; outside influences can lead to health issues and social breakdown of these close-knit societies

Medical plants – rainforest plants are important in helping to treat diseases such as cancer

Figure 1 The importance of the Peruvian Amazon rainforest

How is the Peruvian Amazon under threat from human activity?

The main threats in the Peruvian Amazon are concerned with exploiting the resources of the rainforest for economic gain:

- **Timber** – the many varieties of hardwood trees in the rainforest (for example mahogany) are highly valued for furniture and construction. An estimated 95 per cent of the logging is unregulated and illegal. **Deforestation** often results in vast areas of rainforest being destroyed, causing significant harm to the ecosystem.

> **Key term**
>
> **Deforestation**: the cutting down of trees, transforming a forest into cleared land for other uses such as building or growing crops

- **Energy** – there are large reserves of oil and gas in the rainforest. China has invested in oil extraction in the Madre de Dois region, home to over 10 per cent of the world's bird species. Oil spills can damage ecosystems and polluted rivers. There are plans to construct up to fifteen large dams in the Peruvian Amazon, primarily to export electricity to Brazil to support its industrial development.
- **Mining** – gold is mined from alluvium in rivers, causing enormous environmental damage and leading to pollution by toxic chemicals, including mercury.
- **Roads** – the Trans-Oceanic Highway is a major development planned to connect a major Brazilian highway to the Pacific ports in Peru. This could lead to significant rainforest destruction and may worsen the illegal felling of trees for timber. Other routes are also planned to develop highways through the rainforest.
- **Agriculture** – lowland areas are being cleared to make way for ranching and the cultivation of commercial crops such as soybean. Fires used to clear land may burn out of control and can lead to widespread habitat destruction.

Now test yourself and exam practice answers at www.hoddereducation.co.uk/myrevisionnotes

How can the Peruvian Amazon be sustainably managed?

The Peruvian government is working with non-governmental organisations (NGOs), such as the World Wide Fund for Nature (WWF), to encourage conservation through sustainable management.

- Since 2000, rainforest harvesting activities (such as logging) require a management plan in line with guidelines from the Forest Stewardship Council (FSC). However, the vast area involved makes such policies hard to implement.
- Reserves have been created to protect indigenous tribes and give them land rights over mineral extraction and the use of their land.
- National Parks and National Reserves have been created to protect ecologically valuable areas.
- In southeast Peru, the Purus – Manu Conservation Corridor (Figure 2) comprises national parks and several reserves for indigenous tribes. With an area of 10 million hectares, it is the largest protected area in the Peruvian Amazon. In 2015, a sustainable management plan was drawn up for the area to restrict economic development, conserve habitats and provide protection for the indigenous people.

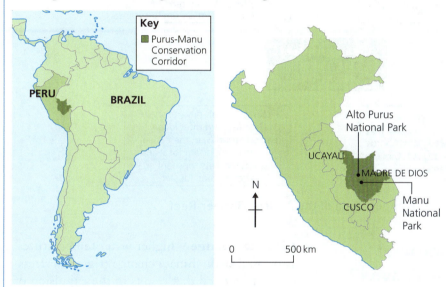

Figure 2 Location of the Purus – Manu Conservation Corridor, Peru

Now test yourself

TESTED ☐

1 Suggest how each of the following is a threat to the Peruvian Amazon rainforest:
   (a) logging
   (b) mining
   (c) agriculture.
2 What is the Purus – Manu Conservation Corridor and what is its purpose?

Answers online

Case study: coral reef under threat

REVISED

Case study: the Andros Barrier Reef, Bahamas

What is the importance of the Andros Barrier Reef?

The Andros Barrier Reef is part of an extensive coral reef in the Bahamas, off the southeast coast of Florida, USA. The reef extends for about 200 km and is centred on Andros Island. In common with other coral reefs, it is teeming with wildlife including over 160 recorded species of fish and numerous species of coral. It is currently one of the healthiest reefs in the world. Figure 3 outlines the reef's importance.

Scientific research – as one of the world's healthiest coral reefs, the Andros Reef is extensively used for scientific monitoring and research

Coastal protection – barrier reefs help to protect coastlines from powerful storm surges associated with hurricanes and tsunami caused by earthquakes; the Andros Barrier Reef helps to protect the Bahamas from hurricanes

The importance of the Andros Barrier Reef

Tourism – many people choose to visit coral reefs for diving and snorkelling as well as deep sea fishing; the Andros Reef accounts for over US$150 million of tourism revenue a year

Fishing – corals provide shelter for fish and offer ideal breeding grounds; in the Bahamas, there are important local and export markets for snapper, grouper and lobster

Figure 3 The importance of the Andros Barrier Reef

How is the Andros Barrier Reef under threat from human activity?

There are several potential threats:
- **Over-fishing** – intensive fishing can deplete the fish stocks and cause an imbalance in the ecosystem. Trawling nets, anchors and outboard motors can cause physical damage to delicate coral organisms. The commercial harvesting of sponges from the Andros Barrier Reef can affect the ecosystem balance.
- **Pollution** – the Andros Barrier Reef is under threat from agricultural chemicals and silt discharged from rivers as it can reduce sunlight penetration and inhibit coral feeding. Oil and chemical discharges from ships also cause harm to corals and fish.
- **Climate change** – higher water temperatures associated with climate change triggers a stress reaction in coral that results in the expulsion of zooxanthellae and 'bleaching' of the coral. This has a huge impact on the coral reef nutrient cycle and eventually the coral dies. In the Andros Barrier Reef, bleaching is expected to be an issue from 2040 according to climate change models.

Now test yourself

1 Suggest how over-fishing is a threat to the Andros Barrier Reef.
2 What causes coral bleaching and why is it a threat to the Andros Barrier Reef?

Answers online

TESTED

How can the Andros Barrier Reef be sustainably managed?

Management of the Andros Barrier Reef is shared between the Department of Marine Resources and the Bahamas National Trust. The government is committed to protect 20 per cent of its near-shore habitat by 2020. As Figure 4 shows, several areas have been designated with protection status:

- Established in 2002, the Andros West Side National Park affords protection from development for over 50 per cent of Andros Island. Management planning aims to balance economic activities such as sponge fishing with conservation and tourism.
- On the east side of Andros Island, the North Marine Park and South Marine Park have regulated fishing, anchoring and scuba diving.
- The Crab Replenishment Reserve seeks to ensure a sustainable crab population for the future.
- In the Exuma Cays Land and Sea Park a coral nursery has been established, enabling threatened species to be conserved and subsequently re-introduced to the coral reef.

Figure 4 Reef management and conservation at Andros Island, the Bahamas

Now test yourself

TESTED

Outline **two** sustainable management schemes aimed at conserving the Andros Barrier Reef.

Answers online

Exam practice

1 With reference to a named tropical rainforest **or** coral reef, explain why biodiverse ecosystems are at risk from human activity. [6]
2 With reference to a named tropical rainforest **or** coral reef, discuss approaches to manage biodiverse ecosystems sustainably. [6]

ONLINE

Exam tip

For both questions, you should select the ecosystem that best suits the question and your depth of knowledge and understanding. In Question 1, make sure that you make clear connections between the human activity and the impact of ecosystem biodiversity. In Question 2, make sure that you stress how each approach offers sustainable (long-lasting, minimal environmental impact) management.

5 People of the planet

The world developing unevenly

What is development?

REVISED

The term **development** can be used to describe the progress of a country as it becomes more economically and technologically advanced. It can also be applied to improvements in people's quality of life – educational opportunities, increased incomes, human rights and healthy living conditions.

> **Key terms**
>
> **Social development**: Improvements in people's quality of life, e.g. literacy, health care and life expectancy
>
> **Economic development**: Improvements in wealth, e.g. GNI and GD
>
> **Environmental development**: Improvements in the quality of the natural world, e.g. air pollution and water quality
>
> **Sustainable development**: Meeting the needs of the present while protecting the needs of those in the future, e.g. the development of renewable energy

How are countries classified?

REVISED

Countries have been classified by global organisations such as the World Bank (WB), the United Nations Development Programme (UNDP) and the International Monetary Fund (IMF). A range of economic and social indicators are used to split the world up into groups that are broadly similar.

The OCR GCSE and A level specifications use the IMF classification.

> **Key terms**
>
> **Advanced Countries (ACs)**: well-developed financial markets, diversified economic structure with rapidly growing service sector, e.g. UK, USA, Japan, Australia
>
> **Emerging and Developing Countries (EDCs)**: do not share all the characteristics required to be an AC but are not eligible for Poverty Reduction and Growth Trust, e.g. South Africa, India, China, Brazil
>
> **Low-income Developing Countries (LIDCs)**: countries eligible for Poverty Reduction and Growth Trust from the IMF, e.g. Nigeria, Bangladesh, Afghanistan

Figure 1 is a development map of the world based upon the International Monetary Fund (IMF) definitions. Notice that most LIDCs are in Africa, with a few in the Middle East, Asia and South America.

Revision activity

Draw a summary table to define each of the IMF development categories and use Figure 1 to include a selection of countries for each category.

Key
- Advanced countries
- Emerging developing countries
- Low-income developing countries

Figure 1 International Monetary Fund (IMF) development classification

How can development be measured?

REVISED

There are several economic and social measures of development.

- **Economic measures** – these are to do with money and include **Gross Domestic Product (GDP), Gross National Income (GNI)** and various monetary measures of poverty and standard of living.
- **Social measures** – these are to do with people and include infant mortality, life expectancy, access to doctors and educational attendance and achievement.

Whilst there are significant similarities between the global patterns produced, there are subtle variations and some indicators tend to be more reliable than others.

It is important to remember that measures are averaged for a whole country. There will often be significant inequalities of wealth and social development **within** a country particularly between major cities and remote rural areas. In fact, inequality is a good measure of the lack of development of a country!

Key terms

Gross domestic product (GDP): the total value of the goods and services produced in a country

Gross national income (GNI): measured as GNI per capita; this means the total income divided by the number of people

Advantages and disadvantages of different development indicators

Measure of development	Advantages	Disadvantages
Gross national income (GNI)	<ul><li>When mapped, GNI can show clear patterns and variations between countries</li><li>It can be used to help prioritise aid payments</li><li>It is easy to calculate using official government figures</li></ul>	<ul><li>These average figures can be misleading – a few very wealthy people in a country can distort the figures</li><li>In poorer countries, many people work in farming or in the informal sector, where their income is not taken into account by official GNI records</li><li>Data about income is sensitive and people may not always be honest</li></ul>

→

Measure of development	Advantages	Disadvantages
Human Development Index (HDI)	• In common with GNI, maps can show clear patterns and differences between countries • HDI provides a composite measure of index, including social aspects as well as wealth • It can be used as a measure of improvement following development initiatives	• It only takes into account a selection of measures and doesn't take account of other important indicators • It can hide variations and inequalities that exist with countries • Data from some countries can be unreliable
Internet users	• Useful as it relies upon other infrastructural improvements, so is a proxy measure for electricity, satellite access and disposable incomes	• Does not take account of variations within countries • Can hide inequalities, especially regarding the use of internet by poor people

Figure 2 is a topological map showing US$GNI per capita. Notice that the area of each country is proportional to the country's GNI. Such maps are visually very powerful and show disparities very clearly.

Key term

Human Development Index (HDI): composite measure using data on income, life expectancy and education to calculate an index from 0 to 1

Figure 2 Topological map showing US$GNI per capita

Figure 2 is taken from www.worldmapper.org, which uses a consistent colour scheme in all of the maps it produces. They divide the world into twelve separate regions (North America, South America, Western Europe, Eastern Europe, Northern Africa, Central Africa, South Eastern Africa, Middle East, Southern Asia, Eastern Asia, Japan and Asia Pacific) and use twelve colours that vary in shades to identify territories within regions.

The twelve regions are coloured in order from poorest to richest by the Human Development Index, with shades of dark red to show the poorest regions, going through the rainbow spectrum of orange, yellow, green and blue, with a shade of violet for the best-off regions.

Figure 3 shows global development according to the widely used Human Development Index (HDI).

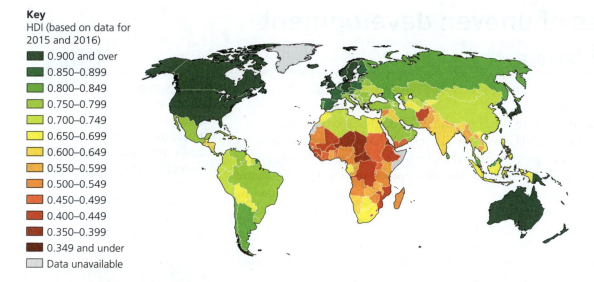

Key
HDI (based on data for 2015 and 2016)

- 0.900 and over
- 0.850–0.899
- 0.800–0.849
- 0.750–0.799
- 0.700–0.749
- 0.650–0.699
- 0.600–0.649
- 0.550–0.599
- 0.500–0.549
- 0.450–0.499
- 0.400–0.449
- 0.350–0.399
- 0.349 and under
- Data unavailable

Figure 3 Global development according to the Human Development Index (HDI)

What are the consequences of uneven development?

- **The 'development gap'** – half of the world's wealth is owned by just 1 per cent of the population, a sign of the huge development gap that exists between rich and poor. Whilst theoretically the LIDCs should be showing signs of improvement, in the last 50 years only 9 countries have moved from LIDC status to the emerging economies category.
- **Health** – low levels of investment in health care and nutrition account for high infant mortality and high birth rates amongst the very poor. These people become trapped in the so-called 'cycle of poverty' from which it is hard to break out.
- **Education** – access to education and improved literacy are vital for people's development and job prospects. This is very patchy in the world's poorest countries.
- **Standards of living** – in LICDs, many people have to endure a lack of clean water and poor sanitation. This can lead to disease and reduce life chances and earning potential, further trapping people in the cycle of poverty.

Now test yourself

1. What are the advantages of internet users as a measure of development?
2. What are the disadvantages of gross national income (GNI) as a measure of development?
3. The Human Development Index is a 'composite' index. What does this mean and why does it make the HDI one of the most widely used measures of development?

Answers online

TESTED

Exam practice

1. What is meant by the term 'development'? [2]
2. Study Figure 1. Describe the pattern of Low-income Developing Countries (LIDCs). [4]
3. Examine how economic and social measures can be used to illustrate the consequences of uneven development. [6]

ONLINE

Exam tip

It is extremely likely that in an exam you will be given a map using social and/or economic measures to show global development. Make full use of the map when answering the question by referring to global regions and specific countries. Use the key to give specific values.

Causes of uneven development

Physical factors

The physical geography of a country or a region can create challenges for economic development.

- **Weather and climate** – heavy rainfall, droughts, extreme heat or cold and vulnerability to tropical cyclones hampers economic development. Vast parts of central and western Africa experience limited and unreliable rainfall. The Philippines and the Caribbean are frequently ravaged by **tropical storms**. In 2016, over 1000 people in Haiti were killed by Hurricane Matthew just six years after 230,000 people were killed by a powerful earthquake.
- **Relief** – mountainous regions, for example countries such as Nepal, tend to be remote and have a poor infrastructure. They are also subject to extreme weather conditions.
- **Landlocked countries** – countries without a coastline lack the benefits of sea trade, which has led to the development of most of the world's most developed nations. A coastline acts as an international border providing huge opportunities for trading with other nations. Eight out of the fifteen lowest ranking countries according to the HDI are landlocked (Figure 1).
- **Tropical environment** – tropical environments (hot and wet) are prone to pests and diseases, which can spread rapidly. Malaria, spread by mosquitoes, and water-borne diseases such as cholera can devastate communities and reduce people's ability to work.
- **Water shortages** – water is essential for life and for development. There are serious shortages of water in some parts of the world, for example in parts of Africa and the Middle East.

> **Key term**
>
> **Tropical storm (hurricane, cyclone, typhoon):** an area of low pressure with winds moving in a spiral around the calm central point called the 'eye' of the storm; winds are powerful and rainfall is heavy

> **Now test yourself**
>
> 1 How can weather and climate affect economic development?
> 2 Study Figure 1.
>   (a) Describe the distribution of landlocked developing countries.
>   (b) Why does the lack of a coastline hinder economic development?
>
> **Answers online**
>
> TESTED

Figure 1 Landlocked developing countries (according to the United Nations)

Human factors

There are several human factors affecting development, including political stability, technology, health care and cultural traditions.

Poverty

The lack of money in a household, community or country slows development. It prevents improvements to living conditions, infrastructure and sanitation, education and training. Without the basics, developments in agriculture and industry will be extremely slow and an economy will simply fail to take off. Figure 2 shows the cycle (spiral) of poverty.

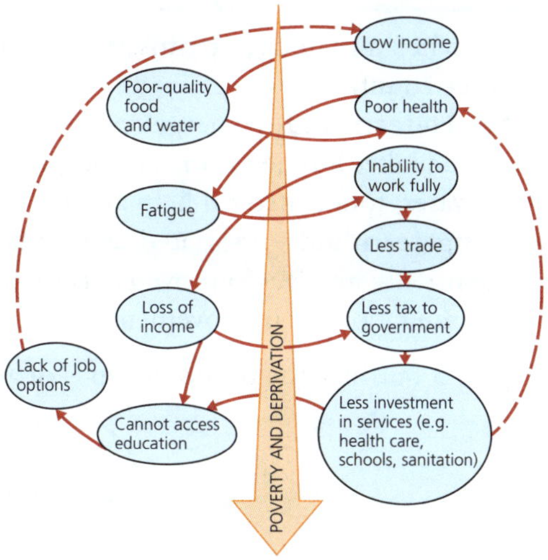

Figure 2 The cycle of poverty

Trade

Trade between nations involves the import and export of goods and services. The vast majority of the world's trade involves the richer countries of Europe, Asia and North America. Most of the world's powerful international companies (TNCs) are based in the ACs. The poorer countries (LIDCs) have limited access to the markets. They have traditionally traded relatively low-value raw materials such as agricultural products or minerals rather than higher-value processed goods. The value of these raw materials (commodities) has fluctuated wildly, causing great uncertainty and instability as countries strive to become developed.

History

Many ACs have experienced a long history of development based upon agricultural and industrial growth and international trading. This has enabled them to become highly developed and relatively wealthy. In recent decades, rapid **industrialisation** has taken place in EDCs such as China, Malaysia and South Korea. The LIDCs have yet to experience significant economic growth.

Colonisation and exploitation of resources

Many LIDCs were colonised by powerful trading nations such as the UK, France, Spain and Portugal. They were exploited for their raw materials and made use of cheap labour. Over 10 million people were exported from Africa to North America to work as slaves. It was during this colonial era that global development became uneven. Most colonial countries became independent in the mid-twentieth century, for example India became independent from the UK in 1947 and Nigeria in 1960. They face huge challenges including poor infrastructure, lack of administrative experience and political instability.

Exploitation of natural resources

Raw materials such as agricultural products (for example cocoa) and minerals (for example copper) were exploited by colonial powers and exported to ACs to further their industrial development. LIDCs were paid low prices for the resources with most of the value adding (processing) taking place in the ACs. This exploitation contributed significantly to uneven development and, to some extent, still continues today.

> **Key term**
>
> **Industrialisation**: the process whereby factories, industry and manufacturing increase and dominate

> **Now test yourself**
>
> 1 With reference to Figure 2, explain why 'poverty leads to poverty'.
> 2 How has colonialism hindered economic development in many LIDCs?
>
> **Answers online**
>
> TESTED

> **Revision activity**
>
> Create a summary spider diagram to identify the main physical and human causes of uneven global development.

How can aid promote and hinder development?

Aid is when a country, organisation or individual gives resources to another country.

- **Short-term aid** – emergency aid usually follows a disaster, such as an earthquake, hurricane or famine. It commonly involves the provision of water, food, medical help and shelter.
- **Long-term aid** – associated with sustainable development, this commonly involves improvements to infrastructure (such as piped water and sanitation) or improvements to education, agriculture and health care.

Figure 3 outlines the different types of aid.

> **Key term**
>
> **Aid**: a transfer of resources from one country to another, typically from a more economically developed country to a less economically developed country

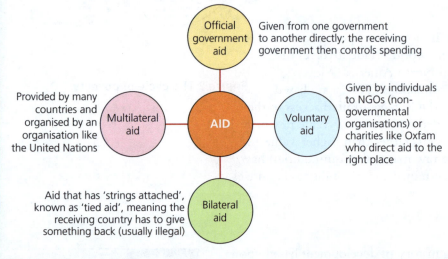

Figure 3 Types of aid

How can aid promote development?	How can aid hinder development?
• The provision of basic services (e.g. clean water) can lead to a healthier life, enabling people to work more effectively and earn money to improve their lives and those of their family • In agriculture, improved knowledge and access to loans can increase food productivity and wealth • Improved access to education will increase literacy rates and standards of living • Improved health care will reduce child mortality, lower birth rates and increase life expectancy	• Inappropriate aid (e.g. machinery) can increase dependency on ACs, e.g. for spare parts and energy • Provision of emergency aid can undercut local food producers, forcing farmers out of business • Lack of coordination between aid donors can result in an imbalance of support • Some animal gift-aid schemes have been criticised for increasing livestock numbers in areas already suffering water shortages and desertification

> **Revision activity**
>
> Design a summary diagram to outline how aid can both promote and hinder development. You could consider social, economic and environmental development using different colours for each.

> **Exam tip**
>
> In Question 2, make sure you focus on 'to what extent' when answering. You should be prepared to express and justify your own opinion, weighing up both sides of the argument. In Question 3, make sure that you express your opinion as part of a balanced argument about the benefits and problems associated with aid.

Exam practice

1 Explain why the exploitation of resources from LIDCs can result in uneven development. [4]
2 To what extent is uneven development the result of physical factors? [6]
3 'Aid can cause as many problems as solutions.' Do you agree with this statement? [6]

Case study: Ethiopia: changing economic development

This section uses Ethiopia as a case study for the economic development of an LIDC.

Ethiopia's location and environmental context

Where is Ethiopia?

Ethiopia is located in the centre-east of Africa and is bordered by six other countries (Figure 1).

- It is the continent's tenth-largest country by area and second most populous after Nigeria.
- Ethiopia's landscape varies from the densely vegetated Western Highlands to arid desert in the Eastern Lowlands.
- For decades the country has suffered from periodic **drought** and famine.
- Ethiopia is Africa's oldest independent country (it remained independent throughout the colonial period) and was a founder member of the United Nations.

> **Key term**
>
> **Drought**: a prolonged period of time with unusually low rainfall; droughts occur when there is not enough rainfall to support people or crops

Figure 1 The location of Ethiopia

> **Now test yourself**
>
> Study Figure 1.
> 1 Name three countries that share a border with Ethiopia.
> 2 Describe the location of Ethiopia within Africa.
>
> **Answers online**
>
> TESTED

What is Ethiopia's environmental context?

Landscape

Ethiopia has a varied landscape. The deeply incised Ethiopian Highlands in the west rise to 4500 metres. Whilst these areas are cooler than the lowlands, soils tend to be thin and the landscape is challenging for communications and the use of machinery. Highland plateaus and the lowlands offer better opportunities for farming, although in places they suffer from overgrazing, soil erosion and desertification. In the southeast of Ethiopia is the Ogaden Desert.

Climate

Ethiopia has three distinct climatic regions (Figure 2). Rainfall is unreliable and can lead to prolonged periods of drought, particularly in the eastern lowlands. In the 1980s, Ethiopia suffered from severe drought and famine. Overgrazing and desertification are significant issues in the east. Despite the unreliable rainfall, food production is high, especially in the wetter and cooler western and central regions.

Ecosystems

Reflecting the diversity of landscape and climate, Ethiopia has a number of different ecosystems including mountains, woodlands and wetlands. Extensive tropical savanna grasslands fringe the highlands and deserts and semi-deserts are found on the eastern lowlands. Biodiversity is high, with thousands of species of plants, birds and animals, including several endangered species.

Natural resources

Ethiopia has reserves of gold, oil and gas but these have yet to be fully exploited. Currently minerals – mostly gold – accounts for about 20 per cent of exports by value.

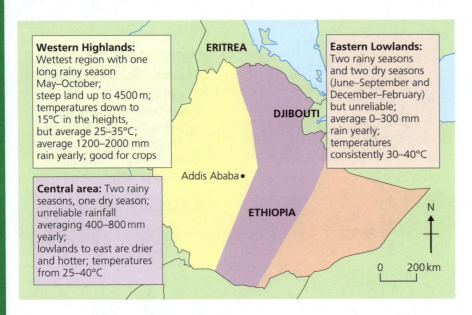

Figure 2 Ethiopia's climatic regions

Now test yourself

TESTED ☐

1 Describe the landscape and climate of the Western Highlands.
2 Describe the climate and ecosystems of the Eastern Lowlands.

Answers online

What are Ethiopia's political links with other states?

REVISED

Date	Links with other states
Pre-1935	Ethiopia was not colonised by European powers during the colonial era.
1935–41	Briefly colonised by Italy.
1941–74	Political unrest and periods of drought prevented development.
1974–87	Military coup supported by the Soviet Union (effectively what is now Russia) and Cuba imposed a communist regime. Many people were killed and lost their land during this period of extreme unrest. An estimated 15 million people were forcibly relocated. Agricultural productivity declined. In 1984–85, famine was triggered by high food prices and drought killed a million people.
1987–2001	Support from other nations, the collapse of the Soviet Union (1991) and Ethiopia's military government resulted in Ethiopia becoming the Federal Democratic Republic (1991). The new government promoted free trade with other nations and provided farmers with cheap fertiliser to enable them to increase food production.
2001+	Long-term aid from countries such as the USA has assisted Ethiopia's development. The government is relatively stable and money has been invested through the Growth and Transformation Plan to improve agricultural productivity through skills training.

Revision activity

Present the information in the table in the form of a timeline.

How has Ethiopia's economy developed?

REVISED

Ethiopia is an LIDC and, with an HDI of just 0.435, it is one of the world's poorest countries. With a GNI of just US$505 per capita (2015), average incomes are significantly lower than the world average of US$10,858 per capita.

Figure 3 shows very slow growth in Ethiopia's wealth since 2004. Ethiopia is also significantly less wealthy than Sub-Saharan Africa and other LIDCs across the world.

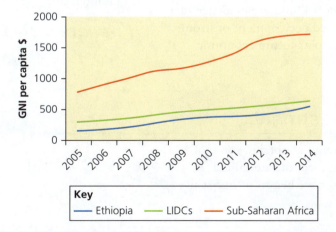

Figure 3 Ethiopia's wealth over time

Ethiopia's international trade: imports and exports

Ethiopia has a trade deficit, in that it exports goods to the value of US$3 billion but imports goods worth US$11 billion. In order to experience economic and social development, this trading deficit needs to be reduced. This will enable the country to spend more money on improvements to infrastructure, education and health care.

Ethiopia's imports

Ethiopia's top five imports are petroleum, trucks, fertilisers, construction and wheat. These imports – petroleum to fuel industry and transport, trucks to improve transport, fertilisers to improve agricultural production and construction equipment for building – indicate that Ethiopia is striving for economic development.

With the exception of wheat, these are all processed, high-value products which explains Ethiopia's trade deficit. It is perhaps ironic that Ethiopia exports agricultural products but still has to import wheat to feed its people!

Ethiopia's exports

Currently, Ethiopia's exports are dominated by agricultural produce, particularly food and flowers. Figure 4 shows Ethiopia's exports in 2015. Coffee dominated Ethiopia's exports up to 2013, from when oilseed has become the most important export. Most of the crops are grown in the wetter and more productive Ethiopian highlands.

Dependence on commodities means that the economy is very vulnerable to factors affecting production, such as weather and climate, and global economics which will affect world prices. Issues with ground transport and storage will also affect the quantity and quality of food exports. Much of what is sold abroad is of low value, with only limited 'value-adding' processing taking place in Ethiopia. These factors of economic insecurity could affect Ethiopia's future development.

Ethiopia's trading partners

Ethiopia has strong trading relations with several countries, including China, India, Saudi Arabia, Germany and Switzerland. China, for example, receives 21 per cent of Ethiopia's exports and provides 11 per cent of the country's imports. These global links are evidence of a stable government and economy. The continued development of strong links with foreign countries will support Ethiopia's future economic development.

Figure 4 Ethiopia's exports, 2015

Total	$5.44 billion
Coffee	17%
Cut flowers	13%
Refined petroleum	11%
Gold	11%
Other vegetables	10%
Other oily seeds	8.7%
Dried legumes	4.3%
Bovine	3.2%
Sheep and goat meat	1.9%
Gas turbines	1.9%
Sheep and goats	1.5%
Other animals	1.1%
Tanned sheep hides	1.1%

Now test yourself

1 Study Figure 4. What are Ethiopia's top five exports?
2 What are the potential problems with an over-dependence on the export of agricultural products?
3 Why is it important for Ethiopia to have strong trading links with foreign countries?

Answers online

TESTED

The role of international investment

Ethiopia has developed strong links with foreign countries, some of whom provide international assistance in the form of foreign aid. A number of trans-national companies (TNCs) have begun to invest in Ethiopia.

Trans-national companies are huge global organisations that operate around the world but usually have their headquarters in rich countries (ACs). There are several TNCs operating in Ethiopia (Figure 5). Most are involved in manufacturing, investing money, materials and expertise that would not readily be available in the country.

Figure 5 TNC investment in Ethiopia

Company	What do they do in Ethiopia?
Hilton Hotels	Leisure and recreation services, hotels
Siemens	Manufacturing of telecommunications, electrical items, medical technology
General Electric (GE)	Aviation manufacturing, delivering rail links
Afriflora	Flower growing, the world's largest producer of fair trade roses
Dow Chemicals	Manufacturing chemicals, plastics and agricultural products
H&M	Textiles manufacturing, university education in textiles

Revision activity

Use a spider diagram to describe some of the advantages and disadvantages of TNCs.

It is possible to identify several advantages and disadvantages of TNCs:

Advantages of TNCs	Disadvantages of TNCs
• Large companies provide employment and training of skills • Modern technology is introduced • Companies often invest in the local area, improving services (e.g. roads, electricity) and social amenities • Local companies may benefit by supplying the TNCs • TNCs have many international business links, helping industry to thrive • The government benefits from export taxes, providing money that can be spent on improving education, health care and services	• TNCs can exploit the low-wage economy and avoid paying local taxes • Working conditions may be poor, with fewer rules and regulations than exist in richer countries • Environmental damage may be caused • Higher paid management jobs are often held by foreign nationals • Most of the profit goes abroad rather than benefiting the host country • Incentives used to attract TNCs could have been spent supporting Ethiopian companies instead

Changes in employment, social factors and technological developments

REVISED

Employment structure

Ethiopia has a large and rapidly growing (2.6 per cent per year) population of over 94 million people. It is the thirteenth most populous nation in the world. Birth rates are beginning to fall and life expectancy is rising.

Ethiopia's economy is dominated by agriculture, which accounts for 80 per cent of employment. Most people are involved in **subsistence** farming, growing crops in the highlands and rearing livestock in the hotter and drier lowlands. Recent years has seen a growth in manufacturing, mostly driven by foreign investment through TNCs. The tertiary sector is growing due to the development of tourism.

Key term

Subsistence: only producing enough goods to meet your own basic needs, with no extra to trade

Education and health care

Access to education and health care has improved as a result of massive government investment and support from overseas.

Access to education	Health care improvements
• 96% of children are enrolled in primary education (up from 50% in 1990) • An increased proportion of girls are enrolled in primary schools (93% compared with just 43% in 2000) • Very few girls continue into secondary education – the society is traditionally very male orientated with girls expected to be home-makers • Education quality varies – the literacy rate is just 36%	• Infant mortality has dropped significantly (97/1000 in 1990 to 45/1000) • 65% of children are vaccinated against preventable diseases • 100% of the population can receive a free malaria net (malaria accounts for 20% of child deaths) • 89% of the population live within 10 km of a doctor

Technology

The state-owned technology company Ethio Telecom operates the country's telecommunications (telephones, internet). The lack of competition has resulted in a patchy network, particularly in the countryside. In 2015, only 4 per cent of the population had internet access and 12 per cent owned mobile phones. There are no credit cards or international banking systems.

Recently, China has invested in mobile phone technology. This is likely to trigger rapid future growth in mobile-phone sales and internet access.

Revision activity

Use a spider diagram to summarise the changes that have taken place in education and health care in Ethiopia.

Aid projects

REVISED

Ethiopia has benefited from international aid and debt relief. Some 5 million people receive food aid each year. Several charities are involved in aid projects.

Oxfam's 'goat aid' provides a pair of goats to young girls who are encouraged to breed to create a commercial flock. The goats produce milk that can be made into cheese, improving health. Surplus can be sold and money used to invest in clothes, food and education (Figure 6). This project deliberately targets girls to raise their status and security, and tackle social ills such as prostitution, forced marriages and early pregnancies.

Mission Aviation Fellowship has supported self-help schemes by transporting tools to enable local communities to create basic sanitation facilities, sink wells to obtain water and provide irrigation for farmers. Such schemes have been successful in improving people's quality of life, reducing disease and increasing farm productivity. There can occasionally be conflicts between communities who might be adversely affected by developments, for example water tables may fall if water is abstracted for drinking or irrigation. Schemes may need regular maintenance and require expertise not available in the local community.

Revision activity

Use a diagram similar to Figure 6 to describe the improvements associated with aid from the Mission Aviation Fellowship.

| Pair of goats given to a 12-year-old girl | Goats are bred to create a flock | Milk is used to drink or make cheese; meat can be eaten | Nutrition improves = better health | Surplus is sold; money invested in education, clothing, food | Social status and wealth improve; flock is re-bred | Cycle continues of breeding, selling, investing and educating | Leads to sustainable increase in wealth |

Figure 6 Sustainable Goat Aid

Rostow's model of economic development

In 1960, the American economist Rostow created a model to show how countries progress through different stages of development (Figure 7). Notice that the model consists of a number of stages or steps that ultimately result in the development of an advanced and wealthy economy.

Whilst Rostow's model provides a useful framework for considering economic development, it assumes equal access to resources and a degree of rigidity in terms of progression. In reality, some countries have skipped stages to progress to an advanced economy.

- Most LIDCs are in the first two stages of the model.
- ACs are in stages 4 and 5, experiencing high levels of consumption, with the majority of people employed in the tertiary and quaternary sectors
- Ethiopia is probably in Stage 2 of the model. Whilst it is still a largely traditional society dominated by agriculture, there have been advances in technology and improvements in education and health care. The pre-conditions for take-off are emerging.

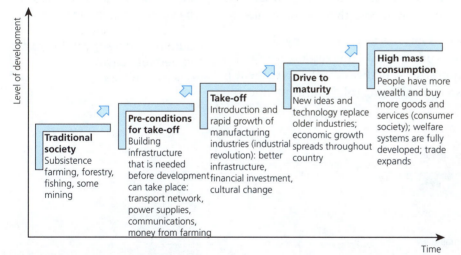

Figure 7 Rostow's model of economic development

> ### Exam tip
>
> Take time to learn Rostow's model so that you can draw a simple sketch. This model is central to a lot of the work on Ethiopia, so you do need to know and understand it.

> ### Exam tip
>
> Question 3 requires you to express and justify your opinion. Use specific facts and figures to support your answer.
>
> When answering Question 4, focus on one aid project and make sure that you refer to both economic (money) and social (people) development. Your answer should be balanced.

> ### Now test yourself
>
> TESTED
>
> 1 Draw a simplified version of Rostow's model of economic development.
> 2 Why is Ethiopia in Stage 2 of Rostow's model?
>
> **Answers online**

The majority of the world's population living in urban areas

In 2007, the UN announced that more than half of the world's population is now living in urban (city) areas. The increase in the numbers of people living in urban areas is called **urbanisation**. Globally, the number of city dwellers increases by an estimated 180,000 people every day. By 2050, 75 per cent of the world's population could be living in towns and cities.

Key term

Urbanisation: the process of towns and cities developing and becoming bigger as their population increases

Cities, megacities and world cities

REVISED

What is a city?

Cities are human settlements, and are near the top of a settlement hierarchy, see Figure 1. As more people live in a settlement, it becomes larger and develops more **functions** and services. Settlements can start off as farms, expand into villages, grow into towns and then become cities.

In the UK, a town becomes a city if its city status is granted by the monarch – there is no other criteria, and towns do not automatically become cities due to their population size. Whether or not a settlement is considered a town or a city can also depend on its role within the local, national or global context.

Key term

Function: a role performed by something; in the case of a city, this may be administrative or related to a sphere of activity

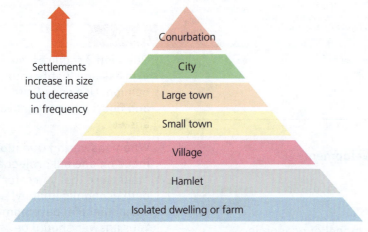

Settlements increase in size but decrease in frequency

Conurbation
City
Large town
Small town
Village
Hamlet
Isolated dwelling or farm

Figure 1 Settlement hierarchy

What is a megacity?

The rapid rate of growth in urban areas has led to the creation of a number of cities with populations of over 10 million people, known as **megacities**. They are often, though not always, capital cities.

> **Key term**
>
> **Megacity**: usually defined as a city that has a population of over 10 million, although the exact number varies.

What is a world city?

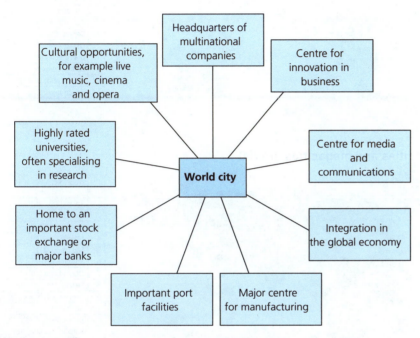

Figure 2 Common characteristics of world cities

> **Key term**
>
> **World city**: a city considered to be an important node in the global economic system and one that has iconic status and buildings, e.g. London and New York

The distribution of megacities

In the 1950s there were only two megacities, but there are now over 30.
Some of the cities have grown to populations of over 20 million people.
Although the number of megacities is growing, they still have only
5 per cent of the world's population at present.

Key

Growth in cities and
megacities (millions)

● 1–5

● 5–10

● 10+ (megacity)

	Top 15 cities in 1950 (millions)
1	New York-Newark, USA (12)
2	Tokyo, Japan (11)
3	London, UK (8)
4	Paris, France (6.5)
5	Shanghai, China (6)
6	Moscow, Russia (5)
7	Buenos Aires, Argentina (5)
8	Chicago, USA (5)
9	Calcutta, India (4.5)
10	Beijing, China (4)
11	Osaka-Kobe, Japan (4)
12	Los Angeles, USA (4)
13	Berlin, Germany (3)
14	Philadelphia, USA (3)
15	Rio de Janeiro, Brazil (3)

Figure 3 Global distribution of major cities and megacities in 1950

Key

Growth in cities and
megacities (millions)

● 1–5

● 5–10

● 10+ (megacity)

	Top 15 cities in 2014 (millions)
1	Tokyo (38)
2	Delhi (25)
3	Shanghai (23)
4	Mexico City (21)
5	São Paulo (21)
6	Mumbai (21)
7	Osaka-Kobe (20)
8	Beijing (19)
9	New York-Newark (19)
10	Cairo (18)
11	Dhaka, Bangladesh (17)
12	Karachi, Pakistan (16)
13	Buenos Aires (15)
14	Calcutta (15)
15	Istanbul, Turkey (14)

Figure 4 Global distribution of major cities and megacities in 2014

Exam practice

1 Use Figure 6 on page 99 to complete the following sentence: 'The
   population of Brazil is predicted to be ... urban in 2030.'   [1]
2 Explain the challenges and opportunities faced by megacities.   [4]
3 Describe the characteristics of a world city, such as London.   [4]
4 Describe the difference between the distribution of megacities in
   1950 and 2014.   [3]

ONLINE

Urban growth rates

The pattern of growth has not been the same everywhere. Figure 5 describes the general patterns of growth in advanced countries (ACs), emerging and developing countries (EDCs) and low-income developing countries (LIDCs).

Figure 5 Growth in ACs, EDCs and LIDCs

ACs	EDCs and LIDCs
<ul><li>Cities in Europe and North America reached the peak of their growth in the 1950s or earlier.</li><li>The most sustained period of growth took place during the Industrial Revolution in the late 1700s to 1800s.</li><li>The 'baby boom' following the Second World War and the building of new houses led to urban sprawl and the growth of cities.</li><li>London and Paris were the first 'millionaire' cities (population of 1 million).</li><li>Most ACs now have populations that are more than 70 per cent urban.</li></ul>	<ul><li>Cities in Asia and Africa have now overtaken the earlier cities of Europe and North America.</li><li>Economic development in urban areas has driven rural–urban migration, causing younger people in rural areas move to urban areas in search of jobs.</li><li>Many of these people then have children in the city, leading to high rates of natural growth.</li><li>Almost 200 million people moved to urban areas between 2000 and 2010.</li><li>Currently, the highest levels of growth are seen in cities such as Dhaka (7%), Lagos (5.6 %) and Delhi (4.6%).</li></ul>

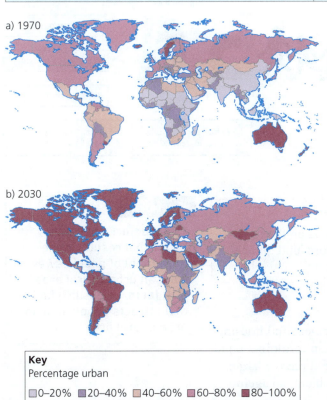

a) 1970

b) 2030

Key
Percentage urban

☐ 0–20% ■ 20–40% ■ 40–60% ■ 60–80% ■ 80–100%

Figure 6 The predicted percentage urban population of countries in 2030

Causes and consequences of rapid urbanisation in LIDCs

Causes of rapid urbanisation in LIDCs

LIDC cities are growing the fastest in the world. The highest population growth rate is in Africa, where urban populations are expected to triple in size by 2050. This increase is due to both **rural to urban migration** and **internal growth**.

Push and pull factors in migration

> **Key terms**
>
> **Rural to urban migration**: people moving from rural areas to live in the cities
>
> **Internal growth**: when people who have moved into the cities have lots of children

Push factors			Pull factors		
Few services, such as education and health care	Wages are at poverty levels in many countries	Lack of job opportunities – jobs tend to be limited to agricultural work	Greater range of employment with higher wages	Better health care systems and schools	Stories filter back to the villages of people doing better in the city, encouraging more to move
Poorer infrastructure					Political and religious freedom
Natural disasters, such as drought and flooding					Stable government
Poor electricity and power supplies					Better access to food
Crop failure					Better-quality housing
Lack of clean water	Increased food insecurity as more young people leave to live in the cities		More transport networks, such as road, rail and air		More entertainment – the 'bright lights' of the city

Push factors   Pull factors

Figure 1 Push and pull factors

> **Exam tip**
>
> Students commonly mix up push and pull factors. Focus on the terms: push factors 'push' people out of the village; pull factors 'pull' people into the city.

> **Key terms**
>
> **Push factor**: a negative factor that results in the movement of people away from an urban/rural area
>
> **Pull factor**: a positive factor that attracts people into an urban/rural area

Natural growth

Once people have arrived in the city and found employment and housing, they tend to have children. This increase in birth rate can result in a rapid rate of population growth, particularly in LIDCs where there is a large, youthful population. ACs tend to have the opposite problem: an ageing population.

Now test yourself

1 What are the two causes of rapid population growth in LIDC cities?
2 What are the environmental push factors from villages?
3 What are the economic pull factors into the cities?

Answers online

Consequences of rapid urban growth in LIDCs

Informal sector

The **informal sector** involves people finding their own employment. This includes jobs as beach vendors, shoe shiners, car washers and litter pickers. It is estimated that over 50 per cent of jobs in Mexico are in the informal sector, and up to 80 per cent of jobs in India.

Jobs in the informal sector require little capital to set up, require few skills, are labour intensive and small scale. People working in this sector do not pay taxes and therefore do not contribute directly to the country's gross national product (GNP). These workers do not have any legal rights and would not receive advantages such as holiday and sick pay.

> **Key term**
>
> **Informal sector**: refers to jobs that don't offer regular contracted hours, salary, pensions or other features of more formal employment; may refer to illegal or unlicensed activity

Informal housing

Informal housing, also known as slums or squatter settlements, are built on land that does not belong to the people building on it. It is usually land that is unsuitable for building, such as:

- a river bed, which can fill with water after rains
- land that is close to industrial activity, which could be bad for people's health
- steep and unstable slopes, which are vulnerable to landslides and flooding
- land either side of railway tracks.

Infrastructure is poor in areas of informal housing, and there are problems with the reliability of electricity and water supplies. Children may have to leave education and work in the informal sector to support their families. With high population densities, disease spreads easily and crime is common. Over time, areas can be improved through self-help schemes.

Figure 2 Slums and skyscrapers in Manila, Philippines

Exam practice

1 Describe the common locations for informal housing in LIDC cities. [3]
2 Explain the challenges faced by people living in informal housing. [4]

ONLINE

Challenges and ways of life in cities in LIDCs and EDCs

Case study: A city in an emerging developing country

Case study: Rosario, Argentina

Location and importance

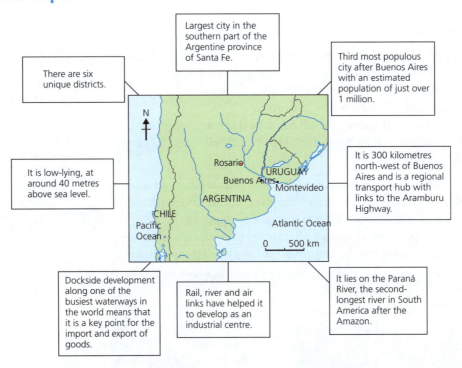

Largest city in the southern part of the Argentine province of Santa Fe.

There are six unique districts.

Third most populous city after Buenos Aires with an estimated population of just over 1 million.

It is low-lying, at around 40 metres above sea level.

It is 300 kilometres north-west of Buenos Aires and is a regional transport hub with links to the Aramburu Highway.

Dockside development along one of the busiest waterways in the world means that it is a key point for the import and export of goods.

Rail, river and air links have helped it to develop as an industrial centre.

It lies on the Paraná River, the second-longest river in South America after the Amazon.

Figure 1 Location map for Rosario, Argentina

Patterns of migration

- The culture of Rosario has been enriched by the fact it has attracted people from across Argentina.
- Arrivals included the Spanish in the sixteenth century and migrants from across Europe in the nineteenth century.
- More recently, there has been an influx of migrants from other countries within South America and as far away and China and Taiwan.
- Italians have formed a significant number of immigrants to Rosario over the last century, which has influenced the city's culture, particularly the food and architecture.
- The city has a young demographic with a relatively high birth rate, though there are signs of an ageing population across the city.

Way of life

- The city has been described the socialist hub of the country. It has close links with trade unions due to its industrial heritage.
- Residents and visitors have good opportunities for shopping in El Centro, the central mall.
- Argentina has a close cultural connection to meat, with one of the highest meat consumptions per capita in the world. The country is the third largest exporter of meat products. Cattle graze the large grassland areas and the *gauchos* (cowboys) are still important. The *asado*, or grill, is an important feature of many restaurants.
- The Argentinean flag comes from Rosario and was raised in the city for the first time. The National Flag Memorial is located on the bank of the Paraná.

Challenges

- **Housing availability**: there are many slums, which house 100,000 people and occupy 10 per cent of the space in the city. The city's infrastructure cannot keep pace with the new arrivals, available land is used up and slums develop along roads and railways.
- **Transport provision**: the Port of Rosario often becomes clogged up with silt and needs to be dredged in order to allow ships to reach the city. Public transport consists of buses, trolleybuses and taxis. In 2012, new bus lanes were added which can be used for buses and taxis in order to reduce congestion. There was a project to build a high-speed train to join Rosario and Buenos Aires but the project was suspended due to high costs. A metro system was proposed, but they are looking at creating a new urban tramway network instead.
- **Waste management**: due to a growing population, waste management has become a pressing issue, with health and safety and environmental impacts.

- **Unemployment**: there were riots in the city as recently as 2001, with high unemployment rates and economic problems leading people to loot supermarkets. As the employment situation improves, so does the level of crime.
- **Crime**: slum districts, known as *villas miserias*, are beset by high levels of poverty and crime. There are violent drug wars in some districts and criminals have allegedly infiltrated the police and football teams to take control.

Sustainable strategies

In Rosario, food makes up 15–30 per cent of waste in landfills. In order to reduce food waste, there have been:

- partnerships with the food industry to increase awareness of the problem of food waste and encourage responsible consumerism
- working with food banks to encourage food donation
- encouraging separation of food waste, so that it can be treated and turned into compost
- giving incentives to businesses to donate unsold food to food banks, or convert it into food for animals.

Exam practice

With reference to your chosen city case study, examine the challenges faced within **one** LIDC or EDC city. [8]

ONLINE ☐

Exam tip

You only need to know one sustainable strategy, in detail, for each city.

Exam tip

Students often refer to many examples when a question clearly specifies **one**. This is such a common mistake that the number is normally marked in bold to draw your attention to it.

6 Environmental threats to our planet

Climate changes from the start of the Quaternary period

How the climate has changed since the Quaternary period

REVISED

What is the Quaternary period?

The Earth is believed to be 4.55 billion years old. The period of time that stretches from 2.6 million years ago to the present day is called the **Quaternary period**. The entire Quaternary period is often called an **ice age** due to the presence of a permanent ice sheet on Antarctica.

Climate change during the Quaternary period

There has been **climate change** during the Quaternary period. Temperatures have fluctuated wildly, but overall have gradually cooled. There have been cold 'spikes', which are known as glacial episodes. In between each cold spike are warmer **inter-glacial episodes**. Today we live in an inter-glacial episode. The average temperature today is higher than almost all of the Quaternary period, as can be seen in Figure 1.

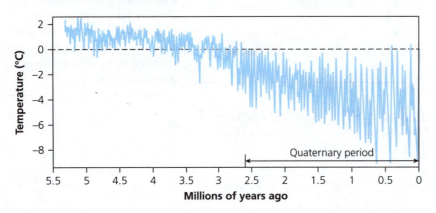

Figure 1 Average global temperatures for the last 5.5 million years

> **Key terms**
>
> **Quaternary period:** the most recent geological period covering the last 2.6 million years, during which time there were several cold and warm periods
>
> **Ice age:** a glacial episode characterised by lower than average global temperatures and during which ice covers more of the Earth's surface
>
> **Climate change:** changes in long-term temperature and precipitation patterns that can either be natural or linked to human activities
>
> **Inter-glacial periods/ episodes:** historic warm periods in between glacial periods where conditions were much the same as they are today

Climate change during the last 400,000 years

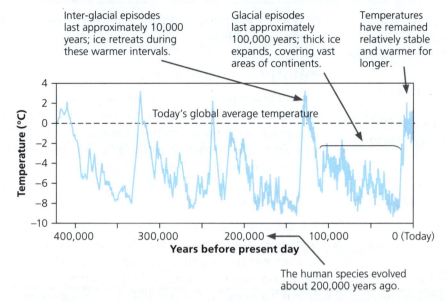

Inter-glacial episodes last approximately 10,000 years; ice retreats during these warmer intervals.

Glacial episodes last approximately 100,000 years; thick ice expands, covering vast areas of continents.

Temperatures have remained relatively stable and warmer for longer.

Today's global average temperature

The human species evolved about 200,000 years ago.

Figure 2 Trends in average global temperatures (400,000 years ago to the present day)

Climate change since 1000AD

REVISED

There have been several periods of warming and cooling since 1000AD. The Medieval Warming Period lasted from 950 to 1250AD. In some regions, the temperatures were equal or higher than today. However, overall temperatures were lower on average than today.

The Medieval Warming Period was followed by the much cooler 'Little Ice Age' from 1300 to 1870. Europe and North America experienced much colder winters than today. Rivers and seas around the UK froze.

Modern warming

In comparison to average temperatures from 1901–2000, average global temperatures have increased in the last few decades.

Now test yourself

TESTED

1 When did the Quaternary period start?
2 How has temperature changed during the Quaternary period?
3 What is different about the inter-glacial episode that we live in today compared to other inter-glacial episodes?
4 How does today's climate compare to that in the Medieval Warming Period and the Little Ice Age?

Answers online

Evidence for climate change

Since 1914, the Met Office has recorded reliable climate change data using weather stations, satellites, weather balloons, radar and ocean buoys. They have collected the following evidence:

- an increase in average surface air temperature by 1 °C over the last 100 years
- the warmest ocean temperatures since 1850
- an average rise in sea levels of 20 cm since 1900.

Evidence for past climate change can be gathered from a range of sources, for example global temperature data, ice cores, tree rings and painting and diaries.

Global temperature data

NASA use over 1000 ground weather stations and satellite information to map global temperature. Average global temperatures have increased by 0.6 °C since 1950 and 0.85 °C since 1880. However, weather stations are not evenly distributed, especially in Africa, so reliability could be questioned. Computer programmes used to produce global temperature maps do not necessarily make them reliable. Data only goes back to 1880.

Ice cores

Oxygen, carbon dioxide and methane in ice cores can help estimate past temperature (spanning 800,000 years) by comparing it to present levels. Scientists drill deep into the ice in the Antarctic and Greenland to extract ice that is thousands of years old. The data is considered very reliable.

Figure 3 An ice core extracted from the Antarctic ice sheet

Tree rings

Each tree ring represents a year of growth.
- Narrow rings indicate cooler and drier past climate conditions.

- Wider rings indicate warmer and wetter past climate conditions.

Paintings and diaries

Diaries and written observations can suggest evidence of climate change at the time, such as:
- price increases in grain in Europe
- sea ice preventing ships from landing in Iceland
- people emigrating due to crop failures
- winter 'Frost Fairs' held on the frozen River Thames.

Several artists captured much colder winter landscapes in Europe and North America in the seventeenth century than we experience today. Cave paintings of animals in France and Spain drawn between 11,000 and 40,000 years ago show significant climate change. However, it is difficult to date cave paintings accurately, and personal accounts and art are subjective viewpoints.

Now test yourself

1 List the different ways that evidence about climate change is collected.
2 How do tree rings help estimate past temperature?
3 How do paintings show evidence of climate change?
4 Give one problem with using global temperature data as evidence for climate change.

Answers online

Exam practice

1 Using Figure 2 on page 105, describe the pattern of temperature change in the last 400,000 years. [4]
2 Explain how ice cores provide evidence of climate change. [4]
3 Describe the advantages and disadvantages of using paintings and diaries as evidence of climate change. [4]

Exam tip

Circle any plurals in exam questions. This will help you to notice when you need to consider more than one factor. Full marks cannot be gained unless you have obeyed this in your answer.

The causes of climate change

There is evidence that climate change occurred before humans existed. This means that climate change must be a natural phenomenon. However, natural causes alone cannot account for the unprecedented temperature increase since the 1970s.

Natural causes of climate change

Variations in energy from the Sun

Sunspots are darker patches on the Sun's surface. They are caused by magnetic activity inside the Sun. Sunspots increase from a minimum number to a maximum number in a sunspot cycle of about every eleven years. Scientists suggest that the more sunspots there are, the more heat is given off by the Sun. However, solar output from the Sun has barely changed in the last 50 years, so it cannot be responsible for the climate change seen since the 1970s.

Changes in the Earth's orbit

The distribution of the Sun's energy on the Earth varies due to changes in the Earth's orbit. The cyclical time periods that relate to the Earth's orbital changes around the Sun are called **Milankovitch cycles**. There are three of them: axial title, precession and eccentricity.

Axial tilt

The Earth spins on its tilted axis. The angle of the tilt changes due to the gravitational pull of the Moon. When the angle of the tilt is greater, it is associated with a higher average temperature. The angle of the tilt moves back and forth every 41,000 years.

Precession

The Earth is not a perfect sphere; as the Earth spins it wobbles on its axis in a 26,000-year cycle.

Eccentricity

The Earth's orbit around the Sun is not fixed and changes over time from being almost circular to being mildly elliptical. The cycle takes 100,000 years. Colder periods occur when the Earth's orbit is more circular and warmer periods when it is more elliptical.

Volcanic activity

Volcanic eruptions throw huge quantities of ash, gases and liquids into the atmosphere. When sulphur dioxide mixes with water vapour it becomes a volcanic aerosol. Volcanic aerosols reflect sunlight away, which reduces global temperatures. Wind carries material far beyond where it was ejected from the volcano, so the reduced temperatures are also experienced elsewhere.

Key terms

Sunspot: a spot or dark patch that appears from time to time on the surface of the Sun; associated with an outburst of energy from the Sun

Milankovitch cycles: the cyclical time periods that relate to the Earth's orbital changes around the Sun

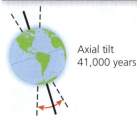

Axial tilt
41,000 years

Figure 1 Axial tilt

Precession
26,000 years

Figure 2 Precession

Figure 3 Eccentricity

Revision activity

Try explaining how the Earth's orbit causes climate change to someone using only actions and no words!

Now test yourself

1 Identify three natural factors causing climate change.
2 What is in a volcanic eruption that reduces global temperatures?
3 Why can volcanic eruptions cause lower temperatures in other regions away from the volcano?
4 What takes:
   (a) 11 years   (b) 26,000 years   (c) 41,000 years   (d) 100,000 years?

Answers online

How human activity contributes to global warming

What is the greenhouse effect?

The **natural greenhouse effect** is a naturally occurring phenomenon that keeps the Earth warm enough for life to exist. Without it, the Earth would be about 33°C colder. This is the process:

1 The Sun's infrared heat rays enter the Earth's atmosphere.
2 The heat is reflected from the Earth's surface.
3 The natural layer of greenhouse gases allows some heat to be reflected out of the Earth's atmosphere but some is trapped. This keeps the Earth warm enough.

The enhanced greenhouse effect

Scientists have proved that natural causes are responsible for climate change, yet they cannot account for the increases in temperature since the 1970s (Figure 6). Humans must also be contributing to the greenhouse effect.

Human activity has increased the layer of greenhouse gases that exits naturally. The thicker layer of greenhouse gases (77 per cent carbon dioxide, 14 per cent methane, 8 per cent nitrous oxides, 1 per cent CFCs) means that less of the Sun's energy is able to escape the Earth's atmosphere, so the temperature increases even more. This is the **enhanced greenhouse effect**.

> **Key terms**
>
> **Global warming:** a trend associated with climate change involving a warming trend (0.85°C since 1880)
>
> **Natural greenhouse effect:** the natural warming of the planet as some of the heat reflected by the Earth is absorbed by liquids and gases in the atmosphere, such as carbon dioxide
>
> **Enhanced greenhouse effect:** the exaggerated warming of the atmosphere caused by human activities, resulting in the natural greenhouse effect becoming more effective

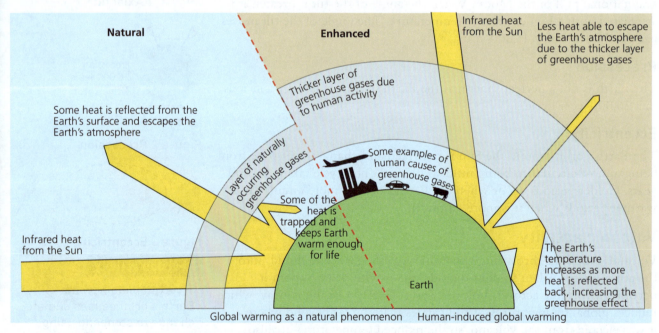

Figure 4 The greenhouse effect: natural and enhanced

Humans enhance the greenhouse effect by generating more greenhouse gases through activities such as those shown in Figure 5.

Figure 5 How human activities contribute to the main greenhouse gases

Greenhouse gas	Human activities that contribute to the enhanced greenhouse effect
Carbon dioxide (CO_2)	● Burning fossil fuels (coal, oil and gas) ● Deforestation (cutting down trees and burning wood) ● Industrial processes (e.g. making cement)
Methane (CH_4)	● Emitted from livestock and rice cultivation ● Decay of organic waste in landfill sites
Nitrous oxides (NO_x)	● Vehicle exhausts ● Agriculture and industrial processes

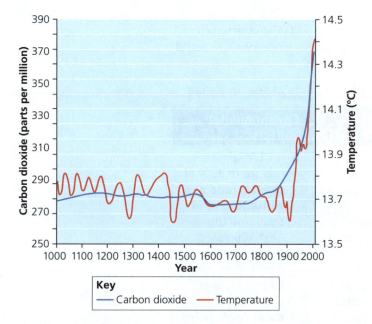

Key
— Carbon dioxide — Temperature

Figure 6 Carbon dioxide and global temperature change

Now test yourself

1 Name the greenhouse gases.
2 Why does the Earth need the natural greenhouse effect?
3 Identify two human activities thought to be causing climate change.
4 Name the fossil fuels.

Answers online

Exam practice

1 Using Figure 6, describe the relationship between carbon dioxide and temperature [2]
2 Explain the enhanced greenhouse effect. [4]
3 Explain why natural factors cannot be solely responsible for climate change. [6]

ONLINE

Revision activity

Draw a pie chart to show the contribution of each of the greenhouse gases.

Consequences of climate change

The effects of climate change are not certain. They are likely to be unevenly distributed across the world and will depend on the human and physical circumstances of the location, for example low-lying coastal countries will be more vulnerable to effects such as flooding, and poorer countries will be more vulnerable as they have less ability to invest in prediction and protection strategies.

The global impact of climate change

REVISED

Sea level rise

The Intergovernmental Panel on Climate Change (IPCC) reported that the sea level has risen by an average of 20 centimetres since 1900. It could rise by another 26–82 cm by 2100.

Figure 1 Social, economic and environmental impacts of sea level rise

	Social impacts	Economic impacts	Environmental impacts
Sea level rise	• **600 million** live in coastal areas less than 10 metres above sea level • Increase in **environmental refugees** due to flooding (e.g. Tuvalu and Vanuatu) • Job losses in fishing or tourism so have to **learn new skills** • **Migration** and **overcrowding** in low-risk areas due to flooding (e.g. in Asia)	• **Valuable agricultural land** lost to the sea or polluted by seawater (e.g. Bangladesh, Vietnam) • Many **world cities**, including New York and London, could be affected by flooding • **Transport infrastructure** damaged by flood water • **Investment in coastal defences** required as UK's current defences are under increased pressure from sea level rise • Loss of income from **tourism** as beaches eroded or flooded; hotels forced to shut	• IPCC estimates that up to 33% of **coastal land and wetlands** could be lost in the next 100 years • Biodiversity lost due to damage by storms and bleaching in **coral reefs** (e.g. the Great Barrier Reef) • **Mangrove forests**, which form natural barriers to storms, are damaged (e.g. the Pacific Islands) • **Fresh water sources polluted** by salty seawater • **Adélie penguins** on the Antarctic Peninsula may decline as ice retreats

Extreme weather events

Single extreme weather events cannot be linked to long-term climate change. However, scientists suggest the increasing frequency of extreme weather events can be blamed on climate change.

Figure 2 Social, economic and environmental impacts of extreme weather events

	Social impacts	Economic impacts	Environmental impacts
Extreme weather events	• Increased **drought**, affecting farming and water supplies (e.g. California 2012–17) • Increased risk of **diseases** such as skin cancers and heatstroke as temperatures increase • Winter-related **deaths** decrease with milder winters	• Extreme weather increases investment in **prediction and protection** • Flood risk increases **repair and insurance costs** (e.g. damage of $9.7 billion in Pakistan in 2010) • Maize **crop yields** decrease by up to 12% in South America; they will increase in northern Europe but require more irrigation • **Skiing industry** may decline in the Alps as there will be less snow	• **Forests** experience more pests, disease and forest fires (e.g. South East Australia had worst bushfires on record in 2009) • Lower rainfall causes **food shortages** for orang-utans in Borneo and Indonesia • **Flooding** in South Asia increases rice yields

Now test yourself

TESTED ☐

1 How are the effects of climate change expected to be distributed across the globe?
2 Name one social, one economic and one environmental effect of rising sea levels.
3 Name one social, one economic and one environmental effect of extreme weather events.
4 Give one positive and two negative effects of climate change.

Answers online

Revision activity

On a map, label a range of impacts as a result of climate change. Colour code them to indicate whether they are social, economic or environmental impacts.

The impact of climate change on the UK

Climate change has a range of economic, social and environmental impacts in the UK. They vary across the UK and can be negative impacts as well as providing opportunities in some sectors, such as increased tourism.

Impact on weather patterns

Summers are expected to become drier, but winters will receive an increase in rainfall. Some rivers will flood more frequently in winter. Temperatures are set to increase, but increases are expected to be greater in the south of the UK.

Figure 3 Impact of climate change on weather patterns in the UK

Economic	The summer heat will lead to growth in the tourist industry in the Lake District, generating jobs and increased revenue. The Cairngorms ski resorts may be forced to close due to lack of snow. This may reduce revenue.
Environmental	Vegetation and ecosystems will move north. Sitka spruce yield may increase in Scotland. New crops such as peaches and oranges could be cultivated in southern England. Agricultural productivity may increase under warmer conditions but require increased irrigation.
Social	The UK's elderly will be more vulnerable during heatwaves but will suffer fewer cold-related deaths in winter. Heating costs will reduce. Water shortages would be experienced by many by the 2050s.

Impact on seasonal patterns

Spring is expected to arrive earlier and autumn to start later. Precipitation is expected to become even more seasonal.

Figure 4 Impact of climate change on seasonal patterns in the UK

Environmental	Climate change will have a huge environmental impact on the behaviour of wildlife and plant species. Bird migration patterns will shift. Some trees and plants will flower earlier and others later. Wildlife species could struggle to survive if the seasons do not match up with their food supply.

Impact on sea level

The UK is expected to be at an increased risk of coastal flooding due to sea level rise.

Figure 5 Impact of climate change on sea level around the UK

Economic	The Thames Barrier will need expensive upgrading or need to be replaced due to the increased risk of flooding in the Thames Estuary.
	Teesside industries on coastal mudflats will be vulnerable to sea level rise.
	Agricultural land may be lost due to managed retreat.
	The tourism industry could be affected by eroded beaches.
Environmental	Salt marshes may become flooded and eroded; however, managed retreat could create new salt marsh habitats.
Social	Cliff collapse may increase, putting properties at risk.

Now test yourself

TESTED ☐

1 How is rainfall changing in the UK as a result of climate change?
2 How is temperature changing in the UK as a result of climate change?
3 Where are temperatures expected to increase the most in the UK?
4 How are seasons in the UK changing as a result of climate change?
5 Is the risk of coastal flooding expected to increase or decrease as a result of climate change?
6 State **three** examples of changes in industry in the UK as a result of climate change.
7 List two social and two environmental impacts of climate change in the UK.

Answers online

Exam practice

1 Identify a negative and positive environmental impact of climate change in the world. [2]
2 Describe the social and economic impacts of climate change in the UK and around the world. [6]
3 Explain how the environment is affected by climate change in the UK. [6]

ONLINE ☐

The global circulation of the atmosphere

How does it work?

There are three large-scale circular movements of air in each **hemisphere** of the Earth's surface. These circular movements, or 'cells', take air from the Equator and move it towards the poles. The cells have a role to play in creating the **climate zones** on Earth.

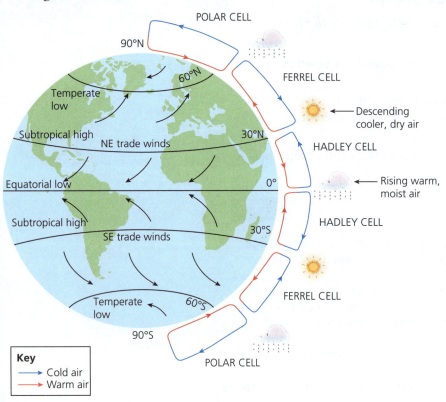

Figure 1 The global circulation system

Circulatory cells

Figure 2 Characteristics of the circulatory cells

	Where is it?	What happens?
Hadley cell	The largest cell, which extends from the Equator to 30° in the north and south.	Winds meet near the Equator and the warm air rises, causing thunderstorms. The drier air then flows out towards 30°, before sinking over subtropical areas.
Ferrel cell	The middle cell, which generally occurs from the edge of the Hadley cell at 30° to 60° in the north and south.	Air in this cell joins the sinking air at the edge of the Hadley cell; it travels across these mid-**latitude** regions until the air rises along the border of cold air with the Polar cell.
Polar cell	The smallest and weakest cell, which occurs from the edge of the Ferrel cell to the poles at 90°.	The air sinks over the higher latitudes at the poles and flows towards the mid-latitudes, where it meets the Ferrel cell and rises.

Key terms

Hemisphere: one half of the Earth, usually divided into northern and southern halves by the Equator

Climate zone: divisions of the Earth's climates into belts, or zones, according to average temperatures and average rainfall; the three major zones are polar, temperate and tropical

Revision activity

Study Figure 1 for one minute, memorising as many details as possible. Draw as much of the model as you can remember on a piece of paper. Look back at Figure 1 to check which details you have missed.

Key term

Latitude: the imaginary lines that surround the Earth ranging from 0° at the Equator to 90° at the poles

High and low pressure

Atmospheric air pressure ranges from low pressure of approximately 980 millibars to high pressure of approximately 1050 millibars. **Low pressure** is created where the two Hadley cells meet and air rises. Where Hadley and Ferrel cells meet, air descends, creating **high pressure**.

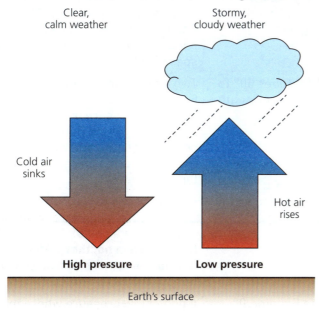

Figure 3 High and low pressure

High pressure

- When air cools it becomes denser and falls towards the ground, leading to high pressure.
- Cool air warms as it reaches the Earth's surface, causing any clouds to evaporate.
- Heavy rain at the Equator means that most of the moisture has gone by the time the air reaches the subtropics.
- High-pressure systems are usually associated with clear skies and dry, hot weather.

Low pressure

- Low pressure causes warm air to rise, after which it cools and condenses to form clouds.
- Moisture falls from the atmosphere as rain, sleet, snow or hail (collectively known as **precipitation**).
- Differences in temperature between day and night are unlikely to be large as the cloud cover reflects solar radiation during the day and traps it at night.

Now test yourself

TESTED

1 At which line of latitude do the Polar and Ferrel cells meet?
2 Would an air pressure of 1036 millibars be high or low pressure?
3 Which circulatory air cell is the smallest?
4 Which circulatory cells meet at the Equator?
5 Why might the climatic conditions be unsettled around 60° latitude in the northern and southern hemispheres?

Answers online

The main climate regions of the world

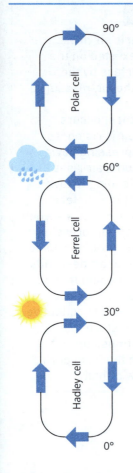

Climate zone	Latitude	Characteristics of the climate
Polar	At the poles 90° north and south of the Equator	Cold air from the Polar cell sinks, producing high pressure. The spin of the Earth creates dry, icy winds. In some parts of **Antarctica**, the average wind speed is 50 mph.
Temperate	Mid-latitudes 50° to 60° north and south of the Equator	Two air cells meet, one warm from the Ferrel cell and one cold from the Polar cell. Low pressure is created as the warm air from the Equator meets the cold air from the poles along a weather **front**. This brings frequent rainfall and is typical of the **UK**.
Subtropical	30° north and south of the Equator	High pressure as a result of sinking air where Hadley and Ferrel cells meet. This creates a belt of deserts including **the Sahara in northern Africa** and **the Namib in southern Africa**. Daytime temperatures can exceed 40 °C.
Tropical	At the Equator, 0° line of latitude	A belt of low pressure where the Hadley cells meet and air rises rapidly. This results in regular heavy rainfall and thunderstorms in places such as **Malaysia in South East Asia** and **northern Brazil in South America**.

Figure 4 Climate zones

Exam tip

You should be able to describe the relationship between the global circulation system and the climate zones. Make sure you refer to precise details such as the cells, places, air pressure and latitude.

Key terms

Temperate: a region or climate with typically mild temperatures
Front: a boundary separating two masses of air with different densities, usually heavier cold air and lighter warm air

Now test yourself

TESTED

1 Which climate zone is found where the Hadley and Ferrel cells meet?
2 Brazil and Malaysia are examples of which climate zone?
3 Why do deserts form at 30° north and south of the Equator?

Answers online

Exam practice

1 Describe two features of the global circulation system. [2]
2 Describe the climatic conditions in a high-pressure belt. [2]
3 Outline the link between Hadley cells and tropical climates. [2]

ONLINE

Extreme weather conditions around the world

Temperature

Coldest place

- **Vostok, Antarctica**: on 21 July 1983, the coldest air temperature ever was recorded at the Russian research station, Vostok: −89.2 °C. It has a height of around 3500 metres, which helps to make it the coldest place on Earth.

Hottest place

- **Al-Aziziyah, Libya**: on 13 September 1922, the world experienced its hottest air temperature ever recorded at 57.8 °C in Libya, which is located 32° north of the Equator.

Precipitation

Driest places

- **Death Valley, USA**: one of the driest places in North America with an average rainfall of 60 mm per year. Storms from the Pacific Ocean travel over a series of mountain ranges before they reach Death Valley, meaning that the moisture has already fallen as rain (Figure 5).
- **Aswan, Egypt**: it has an average rainfall of only 0.861 mm per year; it is close to the Tropic of Cancer.
- **Atacama Desert, South America**: the average annual rainfall is 15 mm. This is due to its location in the rain shadow of the Andes. On its western side the onshore winds do not have enough warmth to pick up moisture from the ocean surface.

Figure 5 Death Valley, USA

Wettest places

- **Mawsynram, India**: this village of 10,000 people copes with an annual average rainfall of 11,871 mm, 80 per cent of which arrives during the seasonal **monsoon** (Figure 6).
- **Ureca, Equatorial Guinea**: located on the southern tip of Bioko Island, Ureca is the wettest place in Africa, with 10,450 mm of rain falling per year.

Figure 6 A sign declaring Mawsynram as the wettest place on Earth

Windiest places

- **Commonwealth Bay, Antarctica**: winds regularly exceed 240 kilometres per hour, with an average annual wind speed of 80 kilometres per hour. Winds carry air from high ground down the slopes by gravity.
- **Wellington, New Zealand**: the strongest gust of wind recorded in Wellington was 248 kilometres per hour. Gusts of wind exceed gale force on 175 days of the year. The mountains either side of Wellington funnel the winds.

> **Key term**
>
> **Monsoon:** heavy rain that arrives as a result of seasonal wind, most notably in southern Asia and India between May and September

> **Exam tip**
>
> You will be required to use specific places and facts to support your arguments. Make sure you are familiar with the map of the world.

> **Revision activity**
>
> Use a blank map of the world to locate the extremes of weather and annotate it with the important information. Do you notice a pattern? You need to be able to link it to your knowledge of the global circulation model and high/low pressure. Are there any anomalies?

Now test yourself

TESTED ☐

1. Where is the hottest place on Earth and what temperature was measured there?
2. Why is rainfall low in the Atacama Desert?
3. How are the winds in Wellington, New Zealand, intensified?

Answers online

Exam practice

The table below lists a selection of the wettest places in the world and their annual rainfall totals.

Location	Rainfall (mm)
Big Bog in Maui, Hawaii	10,272
Debundscha, Cameroon	10,229
Mawsynram, India	11,777
Mount Emei, China	8,169
River Cropp waterfall, New Zealand	11,516
Tutendo, Colombia	11,770

1. Use the table to calculate the:
   (a) median   (b) mean   (c) range of the data.
2. Suggest an appropriate graphical technique to present this data.

ONLINE ☐

Extreme weather conditions causing natural weather hazards

Causes of extreme weather conditions

REVISED ☐

El Niño and La Niña

What causes El Niño?

Scientists continue to study the causes of **El Niño**. It was once thought that sea floor heating following volcanic activity caused it, but this theory is unlikely. A different theory is that small changes in sea surface temperatures cause it. A more likely cause is tropical storms, which trigger the movement of water in a different direction.

Extreme weather conditions

Figure 1 compares normal conditions with the conditions in years when El Niño and La Niña occur.

Figure 1 Comparison of normal weather conditions with El Niño and La Niña years

> **Key term**
>
> **El Niño:** climatic changes affecting the Pacific region every few years; it is characterised by the appearance of unusually warm water around northern Peru and Ecuador, typically in late December; the effects of El Niño include the reversal of wind patterns across the Pacific, causing drought in Australasia and unseasonal heavy rain in South America.

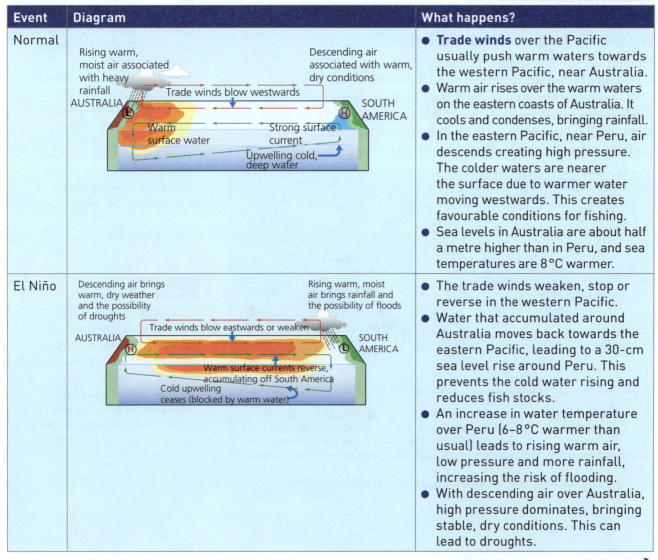

Event	Diagram	What happens?
Normal	Rising warm, moist air associated with heavy rainfall — AUSTRALIA — Trade winds blow westwards — Descending air associated with warm, dry conditions — SOUTH AMERICA — Warm surface water — Strong surface current — Upwelling cold, deep water	• **Trade winds** over the Pacific usually push warm waters towards the western Pacific, near Australia. • Warm air rises over the warm waters on the eastern coasts of Australia. It cools and condenses, bringing rainfall. • In the eastern Pacific, near Peru, air descends creating high pressure. The colder waters are nearer the surface due to warmer water moving westwards. This creates favourable conditions for fishing. • Sea levels in Australia are about half a metre higher than in Peru, and sea temperatures are 8°C warmer.
El Niño	Descending air brings warm, dry weather and the possibility of droughts — AUSTRALIA — Trade winds blow eastwards or weaken — Rising warm, moist air brings rainfall and the possibility of floods — SOUTH AMERICA — Warm surface currents reverse, accumulating off South America — Cold upwelling ceases (blocked by warm water)	• The trade winds weaken, stop or reverse in the western Pacific. • Water that accumulated around Australia moves back towards the eastern Pacific, leading to a 30-cm sea level rise around Peru. This prevents the cold water rising and reduces fish stocks. • An increase in water temperature over Peru (6–8°C warmer than usual) leads to rising warm air, low pressure and more rainfall, increasing the risk of flooding. • With descending air over Australia, high pressure dominates, bringing stable, dry conditions. This can lead to droughts.

→

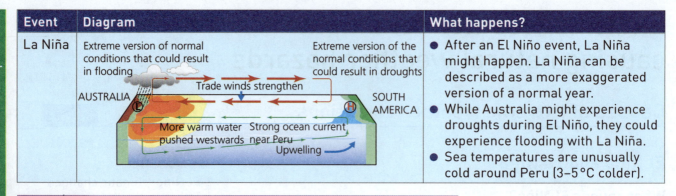

Event	Diagram		What happens?
La Niña	Extreme version of normal conditions that could result in flooding AUSTRALIA Trade winds strengthen More warm water pushed westwards Strong ocean current near Peru Upwelling	Extreme version of the normal conditions that could result in droughts SOUTH AMERICA	● After an El Niño event, La Niña might happen. La Niña can be described as a more exaggerated version of a normal year. ● While Australia might experience droughts during El Niño, they could experience flooding with La Niña. ● Sea temperatures are unusually cold around Peru (3–5 °C colder).

Key term

Trade winds: the prevailing pattern of easterly surface winds found in the tropics, within the lower section of the Earth's atmosphere

The distribution and frequency of tropical storms

REVISED ☐

What are tropical storms?

Tropical storms begin as low-pressure systems in the tropics. They develop into tropical cyclones (also known as hurricanes or typhoons, depending on their geographical location) when wind speeds reach 119 kilometres per hour.

Where do tropical storms occur?

Tropical storms only happen in certain areas:
● typically between 5° and 15° north and south of the Equator
● where the temperature of the surface of the ocean is more than 26.5 °C
● where there is an ocean depth of at least 50–60 metres
● at least 500 kilometres away from the Equator so that the **Coriolis effect** can make the weather system rotate.

Key terms

Tropical storm: an area of low pressure with winds moving in a spiral around a calm central point called the 'eye' of the storm; the winds are powerful and rainfall is heavy

Coriolis effect: the effect of the Earth's rotation on weather patterns and ocean currents, making storms swirl clockwise in the southern hemisphere and anticlockwise in the northern hemisphere

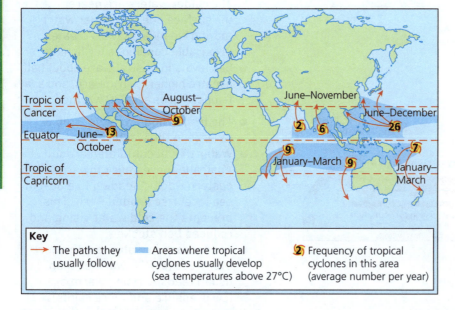

Key
→ The paths they usually follow
▬ Areas where tropical cyclones usually develop (sea temperatures above 27°C)
② Frequency of tropical cyclones in this area (average number per year)

Figure 2 Global distribution and frequency of tropical storms

Exam tip

It is easy to lose marks by giving oversimplified statements, for example stating that 'the ocean water has to be warm' instead of that 'the sea surface temperature has to be 26.5 °C'. Be as precise as you can in your descriptions.

Causes of tropical storms

A number of factors contribute to the development of a tropical storm.

- Temperatures need to cool quickly enough for tall clouds to form through **condensation**.
- The wind speeds need to change slowly with height this is known as wind shear; if the winds in the upper and lower atmosphere are different speeds, the storm will be torn apart.
- Fuelled by warm ocean water, water vapour is rapidly drawn upwards into the low-pressure system; deep clouds rise from the Earth's surface to 15 km.
- The most destruction occurs at the eyewall, where the wind speeds are greatest and rainfall heaviest; this is typically 15–30 km from the centre of the storm.
- When vertical winds reach the top of the **troposphere** at 16 km, they are deflected outwards by the Coriolis effect; this is what makes the storm rotate.

Key terms

Condensation: the process whereby rising water vapour becomes a liquid

Troposphere: an area of the atmosphere, from the Earth's surface to a height of 10–15 km, in which the weather takes place

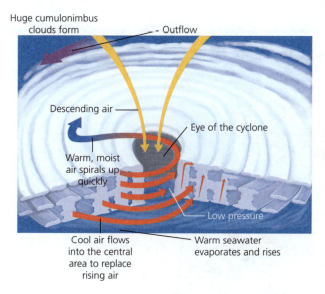

Figure 3 Cross section of a tropical storm

Frequency of tropical storms over time

- There are approximately 80 major tropical storms per year; the most powerful occur in the western Pacific.
- They occur from June to November in the northern hemisphere and November to April in the southern hemisphere.
- The energy released by hurricanes over the last 30 years has increased by 70 per cent.
- During El Niño there tends to be fewer hurricanes in the Atlantic and more tropical cyclones in the eastern part of the South Pacific.
- Scientists disagree about whether climate change has made tropical storms more frequent.

Now test yourself

1 List **five** key features of tropical storms.
2 To what extent are tropical storms increasing in frequency?
3 Draw a diagram to show the formation of a tropical storm.

Answers online

TESTED ☐

The distribution and frequency of droughts

What are droughts?

A **drought** is a prolonged period of time with unusually low rainfall. Droughts occur when there is not enough rainfall to support people or crops.

Where do droughts occur?

- Recent severe droughts have occurred in the Sahel region of Africa, as well as in Middle East in countries that have already been affected by war and conflict.
- Regions that already have an arid (dry) climate are particularly vulnerable if they receive less than their usually very low rainfall. These include Australia, parts of the USA (such as California) and regions of China.
- There are some unexpected examples of drought in the world, such as the Amazon Basin in Brazil, where a drought affected 19 million square kilometres of rainforest between 2002 and 2005.

> **Key term**
>
> **Drought**: a prolonged period of time with unusually low rainfall; there is not enough rainfall to support people or crops

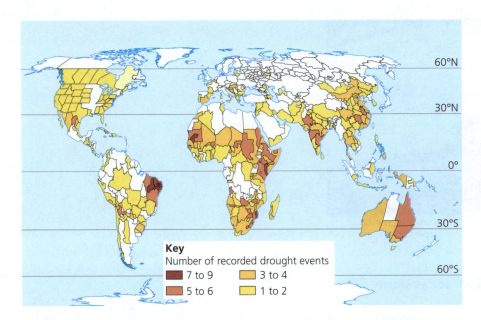

Figure 4 The global distribution and frequency of drought disasters, 1974–2004

Causes of drought

Figure 5 Physical and human factors that can lead to drought

Physical factors	Human factors
• A presence of dry, high-pressure weather systems • El Niño brings descending air and high pressure over Australasia, leading to drought • As global temperatures increase, more water is lost from surfaces through evaporation • The **intertropical convergence zone (ITCZ)** may not move as far north or south as usual, depriving regions, particularly across parts of Africa, of much-needed rainfall	• Excessive irrigation • Deforestation, which reduces transpiration and, therefore, rain • Overgrazing, exposing soils to wind erosion • Dam building, which deprives regions downstream of water • Intensive farming practices

> **Key term**
>
> **Intertropical convergence zone (ITCZ)**: a low-pressure belt that encircles the globe around the Equator; it is where the trade winds from the northeast and southeast meet; as the Earth is tilted on its orbit around the Sun, it causes the ITCZ to migrate between the Tropics of Cancer and Capricorn with the seasons

Frequency of droughts over time

- A 2013 report from NASA predicted that warmer worldwide temperatures will lead to decreased rainfall and more droughts in some parts of the world.
- The Met Office predicts that extreme drought could happen once every decade in the UK in the future.

Droughts can be devastating for people and the environment

Case study: a drought event caused by El Niño and La Niña

Now test yourself

1 Explain how human factors can make the effects of a drought worse.
2 What is the ITCZ and how can it cause a drought?

Answers online

TESTED ☐

REVISED ☐

Revision activity

Create a flashcard for this case study. Create three cards focusing on the keywords from the specification (causes, consequences, responses). Write notes about the case study on one side of the card and key questions on the other side of the card. Practise answering the questions, using the other side of your card to check that you are correct.

Case study: Drought in Australia

Background

From 2002 to 2009, Australia experienced its worst drought for 125 years, which became known as the 'Big Dry'.

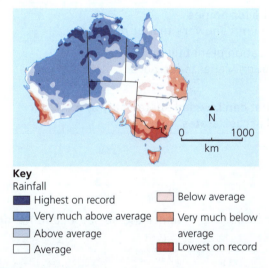

Key
Rainfall
- ■ Highest on record
- ■ Very much above average
- ■ Above average
- ☐ Average
- ■ Below average
- ■ Very much below average
- ■ Lowest on record

Figure 1 The distribution of rainfall in Australia, 1997–2009

Causes

The fact that Australia is often affected by droughts is influenced by a number of factors:
- Australia's geographical location makes it vulnerable to droughts. It is in a subtropical area of the world that experiences dry, sinking air, leading to clear skies and little rain.
- In 2006, the rainfall was 40–60 per cent below normal over most of Australia south of the Tropic of Capricorn.
- When El Niño is in action (see page 119), the chances of rainfall in Australia decreases and it becomes even drier than normal, particularly in eastern Australia.
- The Murray–Darling river basin is home to 2 million people and is under a lot of pressure to supply water to residents and for agricultural production.

Consequences

Figure 2 shows the devastating consequences of the Big Dry.

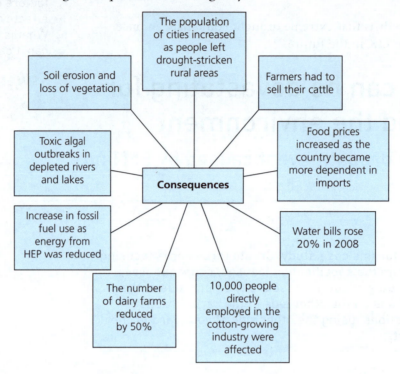

The population of cities increased as people left drought-stricken rural areas

Soil erosion and loss of vegetation

Farmers had to sell their cattle

Toxic algal outbreaks in depleted rivers and lakes

Consequences

Food prices increased as the country became more dependent in imports

Increase in fossil fuel use as energy from HEP was reduced

Water bills rose 20% in 2008

The number of dairy farms reduced by 50%

10,000 people directly employed in the cotton-growing industry were affected

Figure 2 Consequences of the Big Dry

Responses

The different stakeholders found responses to the Big Dry.

Individuals	● Recycling waste water from showers, baths and wash basins (grey water) ● Farmers claiming financial assistance of $400–$600 per fortnight
Local government	● Subsidising rainwater storage tanks for homes ● Legislation to ban car washing and limit showers to four minutes
National government	● A new multimillion-dollar desalinisation plant built in Sydney ● Paying out $1.7 million a day in drought relief to farmers
Scientists and environmentalists	● More efficient irrigation systems ● Calculating the amount of water that can be used sustainably by a state to create a limit that could be traded across states

Figure 3 Responses to the Big Dry

Revision activity

Copy the spider diagram in Figure 2. Use three different colours to indicate the social, economic and environmental consequences. Add a key.

Now test yourself

TESTED

1 Describe one social, one economic and one environmental consequence of drought in Australia.
2 Choose two responses at different scales from Figure 3. Explain how they would help to reduce the effects of the drought.

Answers online

Paper 3 Geographical skills

Paper 3 is made up of two sections:
1 **Section A**: assesses geographical skills and synoptic understanding.
2 **Section B**: assesses your own experience of fieldwork, as well as how you can apply your knowledge to unfamiliar elements of fieldwork enquiry.

A separate twelve-page resource booklet will be provided alongside your exam paper. The resource booklet will contain a variety of geographical information linked to a particular theme. Information could include maps at different scales, diagrams, graphs, statistics, photographs, satellite images, sketches, extracts from published materials and quotes from different interest groups.

Section A: Geographical skills

This section of the paper will consist of a series of questions that are based on a common topic. You will use the resources in the booklet to answer questions on different elements of the topic.

- **Early questions**: these will be worth 1, 2, 3 or 4 marks and will require you to interpret and analyse the information provided in the resource booklet, and will test your geographical skills and your understanding of those skills.
- **Synoptic questions**: these will be longer, 6- or 8-mark questions, which will require you to apply your knowledge and understanding of content from different topics that you have studied in either or both Part 1: Living in the UK and/or Part 2: The world around us. These questions will start with command words like 'Assess' or 'To what extent do you agree?'

The geographical skills that will be assessed are shown in the table below:

Cartographic skills	You need to be able to:	Maps that may be covered:
	<ul><li>select, adapt and construct maps, using appropriate scales and annotations, to present information</li><li>interpret cross-sections and transects</li><li>use and understand coordinates, scale and distance</li><li>extract, interpret, analyse and evaluate information</li><li>use and understand gradient, contour and spot height (on OS and other isoline maps)</li><li>describe, interpret and analyse geo-spatial data presented in a GIS framework.</li></ul>	<ul><li>atlas maps</li><li>OS maps at 1:50,000 and 1:25,000 scales</li><li>base maps</li><li>choropleth maps</li><li>isoline maps</li><li>flow line maps</li><li>desire-line maps</li><li>sphere of influence maps</li><li>thematic maps</li><li>route maps</li><li>sketch maps.</li></ul>

→

Graphical skills	You need to be able to: ● select, adapt and construct appropriate: ● graphs and charts, using appropriate scales ● annotations to present information ● effectively present and communicate data through graphs and charts ● extract, interpret, analyse and evaluate information.	Graphs that may be covered: ● bar graphs – horizontal, vertical and divided ● histograms with equal class interval ● line graphs ● scatter graphs – including best-fit line ● dispersion graphs ● pie charts ● climate graphs ● proportional symbols ● pictograms ● cross-sections ● population pyramids ● radial graphs ● rose charts
Numerical and statistical skills	You need to be able to: ● demonstrate an understanding of number, area and scale ● demonstrate an understanding of the quantitative relationships between units ● understand and correctly use proportion, ratio, magnitude and frequency ● understand and correctly use appropriate measures of central tendency, spread and cumulative frequency including, median, mean, range, quartiles and inter-quartile range, mode and modal class ● calculate and understand percentages (increase and decrease) and percentiles ● design fieldwork data collection sheets and collect data with an understanding of accuracy, sample size and procedures, control groups and reliability ● interpret tables of data ● describe relationships in bivariate data ● sketch trend lines through scatter plots ● draw estimated lines of best fit ● make predictions; interpolate and extrapolate trends from data ● be able to identify weaknesses in statistical presentations of data ● draw and justify conclusions from numerical and statistical data.	
Additional skills	You should also be able to: ● deconstruct, interpret, analyse and evaluate visual images including photographs, cartoons, pictures and diagrams ● analyse written articles from a variety of sources for understanding, interpretation and recognition of bias ● suggest improvements to, issues with or reasons for using maps, graphs, statistical techniques and visual sources, such as photographs and diagrams.	

Exam tip

When studying the resources booklet and answering Paper 3: Section A, make sure you:
● read through the extracts of text carefully
● take time to study the photographs thoroughly
● add notes and annotations to the booklet as you read through it
● look for links and connections between the resources
● consider arguments for and against, or advantages and disadvantages
● write concise and precise answers – don't waffle.

Revision activity

Look back over your notes and the relevant pages of the textbook to ensure that you understand each type of map and graph. Make sure that you feel confident analysing and interpreting the information shown.

Section B: Geographical fieldwork

As part of your GCSE course, you will have completed two geographical enquiries, one physical and one human. Remember that in at least one of the enquiries you must show an understanding of both physical and human geography and their interactions. Make sure that you know which of your investigations involves this.

You will be expected to:
- apply knowledge and understanding to interpret, analyse and evaluate information and issues related to geographical enquiry
- select, adapt and use a variety of skills and techniques to investigate questions and issues, and communicate findings in relation to geographical enquiry.

There are two distinct sets of questions about fieldwork in this section:
- **Individual fieldwork enquiry questions**: these questions are based on **your** two enquiries. The questions will focus on reasons and justification, so you must understand **why** you did things and not just what you did.
- **Generic fieldwork questions**: these are based on the use of fieldwork materials from an unfamiliar context (such as graphs and diagrams) for you to criticise or, for example, information about a location or sample strategy for you to evaluate.

Typical questions about individual fieldwork enquiries

- Explain how the theory behind the investigation determined the data-collection method(s) used.
- Evaluate the effectiveness of your data-collection methods.
- Justify the sampling strategies used in your enquiry.
- Assess the appropriateness of your data-presentation methods.
- Justify the choice of location(s) used to collect data.
- To what extent can the fieldwork results be deemed to be reliable?
- Assess the appropriateness of your data-collection methods.
- Evaluate the accuracy and reliability of your results/conclusions.

Exam tip

When preparing for the fieldwork questions in your exams, make sure you:
- understand the reasons **why** you did things – the choice of location, sampling methods, presentation techniques
- understand the higher-level command words, such as 'justify', 'evaluate', 'discuss' and 'to what extent'
- revise your fieldwork enquiry notes – consider making revision notes
- practise writing concise and precise answers.